Perspectives in Computation

CHICAGO LECTURES IN PHYSICS

Robert M. Wald, *series editor*

Henry J. Frisch, Gene R. Mazenko, Sidney R. Nagel

Other Chicago Lectures in Physics titles available from the University of Chicago Press

Currents and Mesons
J. J. Sakurai (1969)

Mathematical Physics
Robert Geroch (1984)

Useful Optics
Walter T. Welford (1991)

Quantum Field Theory in Curved Spacetime and Black Hole Thermodynamics
Robert M. Wald (1994)

Geometrical Vectors
Gabriel Weinreich (1998)

Electrodynamics
Fulvio Melia (2001)

Perspectives in Computation

Robert Geroch

THE UNIVERSITY OF CHICAGO PRESS
Chicago and London

ROBERT GEROCH is professor emeritus of physics at the University of Chicago. He is the author of *General Relativity from A to B* and *Mathematical Physics*, both published by the University of Chicago Press.

The University of Chicago Press, Chicago 60637
The University of Chicago Press, Ltd., London

Printed in the United States of America

17 16 15 14 13 12 11 10 09 1 2 3 4 5

ISBN-13: 978-0-226-28854-3 (cloth)
ISBN-13: 978-0-226-28855-0 (paper)
ISBN-10: 0-226-28854-4 (cloth)
ISBN-10: 0-226-28855-2 (paper)

Library of Congress Cataloging-in-Publication Data
Geroch, Robert.
Perspectives in computation / by Robert Geroch.
p. cm.
Includes bibliographical references and index.
ISBN-13: 978-0-226-28854-3 (cloth : alk. paper)
ISBN-13: 978-0-226-28855-0 (pbk. : alk. paper)
ISBN-10: 0-226-28854-4 (cloth : alk. paper)
ISBN-10: 0-226-28855-2 (pbk. : alk. paper) 1. Computational complexity. 2. Quantum computers. I. Title.
QA267.7.G47 2009
511.3′52—dc22 2008051322

⊗ The paper used in this publication meets the minimum requirements of the American National Standard for Information Sciences—Permanence of Paper for Printed Library Materials, ANSI Z39.48-1992.

Contents

Chapter 1

Introduction

This book comprises the lecture notes for a course I taught for a small group of graduate students in physics at the University of Chicago in Winter quarter, 2005 and 2007. Our subject was the theory of "computation," broadly defined.

This is a fascinating field. It is full of open questions, many of which look deceptively easy but turn out to be extremely difficult. It is made all the richer by its position on the borderline between mathematics and physics. It is also an extremely active field, with an enormous literature. We shall take a brief walking tour through this subject, trying to pin down a few of the issues that struck me as the most illuminating and interesting. Our emphasis is on what might be called "conceptual issues"—formulating precise definitions, how questions are to be framed, what is possible or not possible in principle, etc.—as opposed to other, more practical, issues. This—the conceptual background of computation—constitutes a coherent subject in its own right; and we focus on it to arrive, quickly and clear-headedly, at some of the central ideas of this field.

This volume should be understandable to readers who know what mathematics is (i.e., what a definition is; what a theorem is; etc.); and, for the last half of the volume, who have a rudimentary knowledge of quantum mechanics. We require virtually no technical knowledge in either mathematics or physics. Thus, the volume should be accessible to many graduate students (and even advanced undergraduates) in mathematics or physics.

The topics that we discuss may be divided into three broad classes: (i) the idea of computability; (ii) the notion of efficiency in the carrying out a computation; and (iii) the possibilities for using quantum mechanics in the computation process. Each of these is an entire field all by itself; and each would require for a proper treatment several courses and several volumes. It is thus well beyond our scope to provide an in-depth survey of even the smallest portion of these subjects.

The first collection of topics—involving the idea of computability—is covered in Sects. 2–8. Sects. 2 and 3 introduce some preliminary notions: the idea of strings over a character set (the currency of the computation process); and the idea of a problem (a "sequence of questions," expressed in terms of such strings). Sects. 4 and 5 introduce the notion of computability—of having an algorithm (expressed as a suitable computer program) to answer each of the questions in succession. It turns out that there exist problems that are not computable. That is, each question in the sequence has a definite answer, but there is no single algorithm for answering them all. In fact, one can write down explicit examples of such problems. (The strategy is to construct the problem by using the computing language itself.) These topics are discussed in Sects. 6 and 7.

The second collection of topics—involving computational efficiency—is covered in Sects. 10–13. Fix a computable problem, together with an algorithm to compute that problem. The key idea (Sect 9) is to introduce a number representing the "difficulty" of that computation, i.e., roughly, the number of steps required to carry it out. The difficulty of a problem as a whole is thus a function of the input string. Sect. 10 contains two remarkable results, due to Blum, which serve to illustrate how computational difficulty works. The first result asserts that there exist essentially "arbitrarily difficult" problems (suitably defined). The second is to the effect that there exist problems for which there is no "best" algorithm: Given any program $\mathcal{P}$ that computes such a problem, there exists another program that computes the same problem, but more efficiently than $\mathcal{P}$ does. This last result seems to destroy any hope of ever assigning, to every problem, an intrinsic difficulty.

This whole idea of computational difficulty suffers from a, potentially serious, drawback. Whereas the notion of computability is essentially language-independent, that of difficulty is not—there are, for example, intrinsically inefficient languages. Better would be a notion of difficulty that attaches to the problem and the method for its solution (but not directly to the language in which that method is written). Sects 11 and 12 represent a proposal to capture such a notion. The strategy is to look for a language that is "as simple as possible, without sacrificing efficiency." It might be interesting to look for a theorem to the effect that this proposed language really does what it is designed to do. Finally, Sect. 13 introduces the idea of incorporating probability into the computation process. It turns out that there is a natural notion of probabilistic programs; of such a program's computing problems; and of the difficulty function of such a program. Probability does not add anything to computability: The probabilistically computable problems turn out to be identical to those computable by a regular program. But—remarkably enough—it remains an open question as to whether the use of probability can effectively reduce the difficulty of a computation.

The third collection of topics—involving the use of quantum mechanics in the computation process—is covered in Sects. 14–21. This is a very active area of research. Rather than trying to summarize this large body of work, we will focus on two—rather narrowly drawn—issues. The first is to translate, into mathematics the physics of utilizing quantum mechanics in the computational process. The second is to examine the issue of whether quantum mechanics can enhance computational efficiency. Sect. 14 is a very brief review of quantum mechanics—for people who already know quantum mechanics. Sect. 15 is a self-contained, four-page exposition of one example, called the Grover construction, in which quantum mechanics seems to hold out the prospect of enhanced efficiency. Those interested only in the role of quantum mechanics in computation may wish to start with this section. Sect. 16 discusses various issues—some of which turn out to be rather subtle—that arise in connection with the Grover construction. These three sections, 14–16, are intended as an informal introduction to the idea of using quantum mechanics as an aid to computation. They serve as motivation for Sects. 17–19, the key sections of this volume. In Sect. 17, we specify precisely what we mean by a "quantum-assisted computation": We introduce a certain precise (mathematical) computer language (a language that involves certain Hilbert spaces and operators on those spaces). This language is designed to reflect what could be done, in the laboratory, using quantum systems. In Sect. 18, we show that, provided matters are set up correctly, quantum-assistance, as here defined, will not enable us to "solve" any otherwise-noncomputable problems. This comes as no surprise (although the "setting up" must be done with some care). In Sect. 19, we define the difficulty function for a quantum-assisted computation.

We now have, for any given problem, the notion of its being computed, and the difficulty in doing so, by a regular program, and also by a quantum-assisted program. We have a "level playing field"—a well-posed mathematical framework within which we may compare quantum-assistance with the lack thereof. There are indications that, for certain problems, quantum mechanics may be utilized in the course of the computation in such a way to enhance the efficiency of that computation. And yet, these indications notwithstanding:

> We have today not one single example of a problem, together with a quantum-assisted computation of that problem, such that we can actually

> *prove* that at least the same efficiency cannot be achieved already without using quantum mechanics.

This is, to my mind, a remarkable state of affairs. The basic obstacle here is, not any uncertainty about how quantum mechanics works in the computation process, but rather the lack of good lower bounds on the difficulty of *regular* computations. Such bounds seem to be lacking for even the simplest problems, e.g., that of multiplying two integers! These issues are discussed in Sects. 20 and 21. In Sect. 20, we prove a theorem to the effect that, if there does exist a problem for which quantum mechanics produces any gain in efficiency, then that gain can never be more than exponential. Finally, Sect. 21 discusses a few ideas about how one might prove (or disprove) that quantum mechanics has the ability to enhance computational efficiency.

I hope that these lecture notes will provide a brief glimpse into this most fascinating field.

Chapter 2

Characters and Strings

Fix a finite set, $\mathcal{C}$, having at least two elements. This $\mathcal{C}$ will be called the *character set* and its elements the *characters*. We shall normally introduce various symbols to denote the various elements of $\mathcal{C}$. For example, $\mathcal{C}$ might have just two elements, and these might be denoted $0, 1$; or $\mathcal{C}$ might consist of 26 elements, with these denoted $a, b, \cdots, z$; or 36 elements, denoted $a, b, \cdots, z, 0, 1, \cdots, 9$. Or, as a final example, $\mathcal{C}$ might consist of 256 elements, denoted by the various ASCII characters.

The underlying choice of character set makes no significant (i.e., no interesting) difference to anything that follows, although a poor choice can turn out to be inconvenient. Eventually (but not right now) we shall allow ourselves to be a little sloppy as to exactly what our character set is.

Fix a character set $\mathcal{C}$. By a *string* over $\mathcal{C}$, we mean a finite, ordered list of characters. Examples of strings for the character sets above are "001010", "unblowupable", "3$dafrq$", and "$\$ = log\ 8+\} - -r\&C$", respectively. The empty list of characters, which we denote $\emptyset$, is also allowed as a string; and it is called the *empty string*. (Thus, to avoid confusion, we shall avoid denoting any character by $\emptyset$.)

The set of strings, over a given character set $\mathcal{C}$, will be denoted $\mathcal{S}$ (or $\mathcal{S}_{\mathcal{C}}$, if there is a chance of confusion as to what the character set is). Note that $\mathcal{S}$ is an infinite set, in contrast to $\mathcal{C}$, which is a finite set. A critical idea in this subject is to pass to "the infinite" in a careful, controlled way. Here, that passage is taking place in the construction of $\mathcal{S}$ from $\mathcal{C}$. This is something that we carry out directly, as opposed to letting $\mathcal{C}$ already be infinite on its own, in any old manner that it chooses.

Part of the reason why the "choice of character set makes no significant difference" is that it is possible to pass from one character set to another. For example, let $\mathcal{C} = \{0, 1\}$ and let $\mathcal{C}'$ be the ASCII character set. Then "writing a byte as eight bits" provides a mapping from $\mathcal{S}_{\mathcal{C}'}$ to $\mathcal{S}_{\mathcal{C}}$. For example, string "$ab$" $\in \mathcal{S}_{\mathcal{C}'}$ might be sent to the string "0000100100001010" $\in \mathcal{S}_{\mathcal{C}}$. Unfortunately, this mapping is not invertible: It is not true that every string over $\mathcal{C}$ arises in this way from *some* string over $\mathcal{C}'$. (Indeed, a string over $\mathcal{C}$ *does* arise in this way if and only if the number of characters of which it consists is divisible by eight.)

Next, fix a character set $\mathcal{C}$, and fix also an ordering of the characters in that set (i.e., a choice of a "first" character, a "second" character, etc. through all the characters in the set). For example, $\mathcal{C}$ might be the lowercase Latin letters (the set of 26 characters, $\{a, b, \cdots, z\}$), and the ordering might be alphabetical. Having made these choices, we now construct an ordering also of the set $\mathcal{S}$ of strings over $\mathcal{C}$, in the following manner: First is the empty string, $\emptyset$; then all the one-character strings, in the ordering of the characters; then all two-character strings, in dictionary ordering; then all three-character strings; etc. So, for instance, in the example just given the ordering of $\mathcal{S}$ is $\emptyset$, "a", $\cdots$, "z", "aa", "ab", $\cdots$, "az", "ba", "bb", $\cdots$, "zz", "aaa", $\cdots$. Now assign, to these strings so ordered, successive integers, beginning with the integer 1. In this way, we set up a correspondence between the set $\mathcal{S}$ and the set $\mathcal{Z}^+$ of positive integers. In short, strings are really just positive integers, in thin disguise. Indeed, we shall allow ourselves to speak of "the nth string," by which we shall mean the nth string in this ordering, where some fixed ordering of the character set $\mathcal{C}$ is implicit (or explicit). When dealing with mathematical issues (such as the manipulation of recursive functions), it is sometimes more convenient to stick entirely with the integers, ignoring character sets and strings altogether. But the strings seem better adapted to dealing with physical issues.

The reason that we required that the character set contain at least two elements is that, for $\mathcal{C}$ having but a single element, the length of the nth string grows linearly with n (rather than logarithmically, when $\mathcal{C}$ contains two or more characters). This behavior would be inconvenient when we come to discuss computational difficulty.

The ordering just given leads to a different way to pass from strings over one character set to those over another. First, order both character sets (in any way whatever), and then identify, for $n = 1, 2, \cdots$, the nth string over the first character set with the nth string over the second. Consider, for example, the character sets given by $\{0, 1\}$ and $\{a, b, \cdots, z\}$, in the indicated orderings. Then the string "1101" over the first character set would be identified, in this manner, with the string "ab" over the second character set. Note that (in contrast with the earlier method) this really is a correspondence between the two sets of strings; that is, this mapping from $\mathcal{S}_{\mathcal{C}}$ to $\mathcal{S}_{\mathcal{C}'}$ is one to one and onto.

Chapter 3

Problems

Fix a character set $\mathcal{C}$. A *problem* is a mapping $\mathcal{S} \xrightarrow{\pi} \mathcal{S}$, that is, a mapping from strings over $\mathcal{C}$ to strings over $\mathcal{C}$. Here is an example of a problem:

> **Example.** Let the character set have 13 elements, $0, 1, \cdots, 9$, y, n, f, and let the mapping π be the following: If the string $S \in \mathcal{S}$ is an integer greater than or equal to two (i.e., if it is not $\emptyset$ or "1", does not contain the characters y, n, or f, and does not have an initial character 0)[1] then let $\pi(S)$ be "y" [yes] if the integer S is prime, and "n" [no] if that integer is not prime. If the string S is not an integer greater than or equal to two, then set $\pi(S) =$ "f" [forget it]. This is indeed a mapping $\mathcal{S} \xrightarrow{\pi} \mathcal{S}$ and so is a problem.

Note that, in this example, we are actually only interested in certain strings (namely, those that represent integers greater than or equal to two). But we cannot confine ourselves

[1] To avoid having to say all this repeatedly, let us agree, here and hereafter, on the following definition: Over any character set that includes the digits $0, 1, \cdots, 9$, a string will be called an *integer* if it contains only those 10 characters, is not $\emptyset$, and (unless it is actually the string 0) does not have initial character 0.

simply to these strings, for, by definition of a "problem," the mapping must apply to *all* strings. So we send the ones we are not interested in to the trash ["f"]. It is convenient to set things up in this way. Suppose, for example, that we had defined a problem to be a mapping from a mere subset of $\mathcal{S}$ to $\mathcal{S}$. Then, for example, it would be false that the composition of two problems is a problem. Even worse, we would have to confront eventually the issue of how we shall determine whether or not a given string is in the "certain subset." The present definition—requiring that the mapping π have domain all of $\mathcal{S}$—puts the burden of sorting all this out back on the mapping π itself (where, as we shall see, it belongs).

In this example, the problem is really the question "Is integer S prime, or is it not?" Note that we could ask essentially this same question by any number of other maps (i.e., by any number of other problems). For instance, we could eliminate y, n, and f from the character set, and then encode the answer as a digit (e.g., 0 for prime, 1 for integer ≥ 2 but not prime, and 2 for not an integer ≥ 2). We could also modify the input. For example, we could let the character set consist of the 26 lowercase Latin letters, impose alphabetical ordering, and let π ask whether, given string S (say, the nth string in the induced ordering on $\mathcal{S}$), the integer n is prime. Thus, we see that a given question may give rise to many problems.

We emphasize that a problem is a *map*, and not the process by which we arrived at that map. Consider, for example, the following problem:

> **Example.** Let the character set again be 0, $1, \cdots, 9, y, n, f$, and let $\pi(S)$ be "f" if S does not represent an integer greater than or equal to two, "y" if it does represent such an integer and that integer is the largest prime, and "n" if it does represent such an integer and that integer is not the largest prime.

This is the same *problem* (i.e., the same map) as that which sends string S to "n" if S represents an integer greater than or equal to two and to "f" otherwise. (This follows, since there is no largest prime.) But this is a very different characterization of the problem than our earlier one. Thus, we see that several apparently different questions may give rise to precisely the same problem.

Think of a problem π as a "broad question," of a particular value for the argument S as a "specific instance" of that broad question, and of $\pi(S)$ as the answer to that question in that instance. Thus, a problem represents the answers to an infinite number of questions (since there is an infinite number of possible input strings). The problem just given of deciding whether an integer is prime illustrates this point. But it is also possible to design a problem that answers a single question. For example, consider.

> **Goldbach's Conjecture.** Every even integer $n \geq 4$ is equal to a sum of two primes.

Let the character set be $0, 1, \cdots, 9, y, n, f$, as before, and let, for each string $S \in \mathcal{S}$, $\pi(S) =$ "y" if Goldbach's conjecture is true and $\pi(S) =$ "n" if Goldbach's conjecture is false. (Note that π does not care what S is. No law says it must.) Since this conjecture is, presumably, either true or false, this is indeed a problem. However, this problem is *either* the problem $\pi'(S) =$ "y" for all S *or* the problem $\pi''(S) =$ "n" for all S. But these are both rather uninteresting problems. Thus, even though the original question ("Is Goldbach's conjecture true?") is interesting, translating it in this way into a problem yields what is guaranteed to be a pretty boring problem. Of course, to know whether π is π' or π'' would be interesting.

Here are some other candidates for problems.

1. Let the character set be $\{0, 1, \cdots, 9, f\}$, and let $\pi(S)$ be "f" if S is not a positive integer and let it be the Sth digit in the decimal expansion of the number π if S is a positive integer. This is a problem.

2. Let the character set be the same, but again let $\pi(S)$ be "f" if S is not a positive integer. But if S is a positive integer, let $\pi(S)$ be a string of the form {digits of a positive integer i}f{digits of a positive integer i'}, where i/i' is within 10^{-S} of the number π. This is not a problem (since the actual mapping has not been specified). Rather, this is a description of a class of problems. There does indeed exist a problem in this class (e.g., expand the number π decimally, as in the first example; stop as soon as you reach a rational within 10^{-S} of the number π; and finally reduce that fraction to lowest terms).

3. Let the character set be $\{0, 1, \cdots, 9, y, n, f\}$. Let $\pi(S)$ be "f" if S is not an integer ≥ 4. If S is an integer ≥ 4, let $\pi(S)$ be "y" if S is a counterexample to the Goldbach conjecture, and let $\pi(S)$ be "n" if it is not. This is a problem. The question of whether Goldbach's conjecture is true is the question of whether or not the problem π is equal to a suitable simple problem (which always answers "f" or "n").

4. Let the character sets consist of the uppercase and lowercase Latin letters, together with an appropriate set of punctuation marks (period, comma, semicolon, question mark, exclamation mark, etc.). Let $\pi(S)$ be "yes!" if S is the Gettysburg address, and "no" otherwise. This is a problem.

5. Let the character set be any ordered set that includes at least the 10 digits (not necessarily in their natural order). Let π send string S over this character set to that integer n that is such that S is the nth string. This is a problem.

It is interesting to note that a progression to ever greater levels of the infinite has taken place. The character set is finite; the set of strings over that character set is countably infinite; and, finally, the set of problems on those strings is uncountably infinite. We pause to give a proof of this last assertion because it illustrates a method, called a diagonal argument, that we shall use several times later. Fix $\mathcal{C}$, and suppose, for contradiction, that we had a countable collection of problems, $\pi_1, \pi_2, \pi_3, \cdots$, that exhausted all problems on this character set. We now introduce a new problem, π, as follows: Say one of the characters is a. On the nth string, S_n, set $\pi(S_n) = \emptyset$ if $\pi_n(S_n) =$ "a"; and $\pi(S_n) =$ "a" otherwise. Then this problem π, so constructed, is not equal to π_n for any n, since by construction $\pi(S_n) \neq \pi_n(S_n)$. Thus, the list $\pi_1, \pi_2, \cdots$ could not have been exhaustive—a contradiction.

Finally, we remark that this notion of a problem is rather robust. For virtually the entirety of the remaining discussion, we shall have before us some problem, as here defined, or other.

Chapter 4

Computability

Fix a character set $\mathcal{C}$. We next wish to introduce the notion of computability of a problem over this character set.

Roughly speaking, a problem $\mathcal{S} \xrightarrow{\pi} \mathcal{S}$ is computable provided there exists a computer that, when run with any given input string S, will ultimately halt, displaying at that point as output precisely the string $\pi(S)$. But what is a "computer"? We cannot take this to mean a physical computer, because no such computer is ever capable of solving any problem. My desktop, for example, has a hard drive with a capacity of only 10 GB. Thus, if I let the character set be, say, ASCII, and let the input string S consist, say, of 10^{11} characters, then surely this computer will be unable run with this input.

More promising would be to consider, rather than a physical computer, a computer language. Consider Fortran.[2] A given Fortran program has no space limitations whatever associated with it. You begin by purchasing some physical computer and running the given program on it. Then if, during the course of the calculation, it turns out that there is insufficient space to complete that calculation, you will be invited

[2] We shall not be concerned here with details of any specific computer languages. In particular, we take "Fortran" as a generic term, which describes languages having such commands as "Set $x = \dots$," "If $(\cdots)$, go to...," "Do, for $I = 1, n\{\cdots\}$ Next," "Print...," "Cat x, y," etc.

to purchase a larger computer and rerun the program on it. So, let us call a problem $\mathcal{S} \xrightarrow{\pi} \mathcal{S}$ "Fortran-computable" if there exists a Fortran program with a single, initial "Input" instruction (allowing the user to input some string S) and a single, final "Print" instruction (allowing the program to display to the user some final string), having the following property: For every choice of the input S, the program ultimately halts (as opposed, e.g., to getting caught in an infinite loop), having printed precisely the string $\pi(S)$.[3] For instance, every example of a problem we have given so far is Fortran-computable in this sense. Indeed, one might even imagine at this point that *every* problem is Fortran-computable.

The difficulty we now face is that there are many computer languages. That is, we also have, defined in a similar way, "C-computable," "Applesoft-computable," etc. But our goal here is to capture by a general definition an abstract notion of "computable," that is, to isolate the general structure of the computing process itself. The danger we face is that the various types of computability, as defined here, will say more about the individual languages that gave rise to them than they do about this general structure. But anyone who is familiar with two or more languages will realize that this difficulty is more one of principle than one of practice. Consider two languages, for example, Fortran and C. You can write a Fortran emulator in C (and, indeed, this is probably what "Fortran" really is!). That is, you can write a C program that will accept as input lines of Fortran code: "Set $x = 7$," etc. The C program will then parse each such Fortran command, figure out what the Fortran language would have done to implement that command, and then itself do precisely that. From the mere existence of such a Fortran emulator in C, it follows that every Fortran-computable problem is also C-computable. (To see this, consider any Fortran-computable

[3] To make this idea into a proper definition, we would have to specify the details of the language "Fortran." We shall not do this, since this entire discussion is intended merely as motivation for what follows.

problem π. Taking the Fortran program that computes this π and applying to it our emulator, we obtain a C program that computes π.) In a similar way, we can write a C emulator in Fortran. We conclude, then, that the Fortran-computable problems are precisely the same as the C-computable problems. A similar argument shows that all the standard languages of the computer world generate precisely the same computable problems.

> **Exercise.** Consider three languages, A, B, and C. Suppose you were given an A emulator in B and a B emulator in C. Could you use these two to construct an A emulator in C? As a second question, consider two languages, A and B. Suppose that the A-computable problems are precisely the same as the B-computable problems. Can you exploit this fact to build an A emulator in B?

How shall we turn this intuitive discussion into mathematics? Lest you imagine that this will be an easy exercise, we now introduce two new "languages."

The first, which we might call "MiniFortran," has just two commands: "Input," which allows the user to input any string S, and "Print $\emptyset$," which causes the computer to print the empty string. There is just one MiniFortran-computable problem, namely that with $\pi(S) = \emptyset$ for every input string S. Clearly, then, there are many fewer MiniFortran-computable problems than Fortran-computable problems. The problem with MiniFortran, of course, is that it is absurdly barren. A language must have a certain degree of richness [essentially, (i) the ability to store plenty of intermediate data, (ii) the ability to manipulate the data, and (iii) the ability to branch, in response to those data] if it is to reach the mainstream (Fortran, C, etc.) of computable problems.

Our second language, "HyperFortran," contains all the commands of Fortran, together with one additional command, with the following structure: "Do, for $I = 1, \infty\{\cdots\}$

Next." Here, "$\{\cdots\}$" consists of various Fortran commands, including those that may reset certain variables. The rule is that the computer will always exit from such a command (i.e., it will never hang here), and on doing so the variables will be set as follows: Consider any one variable, x. If, in the course of the execution of this Do loop, the variable x was set to some value, and did not ever change that assigned value beyond some particular iteration (i.e., beyond some particular I value), then on exit from this command x is to be assigned that value. If, however, x changed its value an infinite number of times during the course of the Do loop, then on exit x is assigned value $\emptyset$. This is, arguably, a legitimate computer command, at least in the sense that it is completely determined what is to be done in response to such a line of code. I grant that HyperFortran seems a little strange at first sight, but, absent a careful definition of the term "language," a reasonable case could, perhaps, be made that it is one. Note, incidentally, that in HyperFortran we can solve Goldbach's conjecture. We would use a program of the following form:

Do, for $I = 1, \infty$
If ($x == \emptyset$ and I is a Goldbach counterexample) set $x = I$
Next
Print x

If Goldbach's conjecture is true, then the empty string will be returned. If it is false, then a counterexample to that conjecture will be returned.

We all know in our hearts that HyperFortran is unacceptable, but it is not so easy to spell out exactly why. True, you cannot run it on a physical computer—but you cannot run Fortran, either; and in any case it is usually a bad idea to try to base mathematics on physical implementability. What HyperFortran does is illustrate that some care is going to be necessary to formulate a suitable definition of "computable."

So, in summary, all "reasonable" computer languages (in some sense we have yet to pin down) seem to produce the same computable problems. Our challenge is to turn this intuitive idea—which is called *Church's Thesis*—into a piece of mathematics. There are at least three different strategies by which one might imagine doing this.

The first strategy begins by producing a formal definition of "reasonable language." This definition would be along the following lines. A "reasonable language" must have some commands. [Presumably, there would be just a finite number of types of commands, but, since arbitrary strings can normally appear in certain commands, there would be within these types an infinite number of actual commands (just like Fortran).] With each command there would be associated something to "do" (such as manipulating a string, going somewhere, etc.). We would require, as part of this definition, that these commands be rich enough to allow one to do "the necessary things for computing" (i.e., store arbitrary amounts of data, manipulate data, input/output, branch), but not so rich that they do "ridiculous things" (such as "Do, for $I = 1, \infty$"). Note that we are *not* specifying any specific language here; rather, we are describing by the definition the characteristics that we will demand of a language in order that we deem it "reasonable." Then given such a language, $\mathcal{L}$, we would call a problem π *$\mathcal{L}$-computable* if there exists a program in $\mathcal{L}$ that, for every input string S, eventually halts, returning exactly $\pi(S)$. Finally, (the crowning result of this strategy) we would prove the following theorem: For $\mathcal{L}$ and $\mathcal{L}'$ any two reasonable languages as defined above, the $\mathcal{L}$-computable problems are precisely the same as the $\mathcal{L}'$-computable problems. The key to this strategy, of course, is discovering the right definition: It has to look simple and not contrived, and at the same time be just right so that the theorem really is a theorem. I feel that it might be enlightening to carry out this strategy—but it looks like a lot of work.

The second strategy begins by noting that, since strings can be replaced by integers, each problem thereby becomes an integer-valued function on integers. We would now introduce some axioms that are intended to characterize the "computable functions." There might be a few simple ones, such as "Every constant function is computable." and "The composition of two computable functions is computable." But then there would be some more complicated ones, requiring that certain constructions involving computable functions result in computable functions. Then, a function would be deemed computable if it arises from these axioms. This strategy has in fact been carried out: It is the subject called recursive function theory. [Recursive functions are precisely the computable (to be defined shortly) problems.] This strikes me as an elegant and promising approach. Its downside is that recursive function theory does not seem especially well matched to physics—and, in particular, not to quantum mechanics. Furthermore, the subject of computational difficulty does not, as far as I am aware, fit in naturally with this strategy.

The third strategy consists of inventing the "simplest possible language" that is still (barely) rich enough that it generates the same computable problems as the real-world languages. We then take computability to mean computability in this language. This is the strategy we shall pursue, in the following section.

Chapter 5

Turing Machines

Fix a character set $\mathcal{C}$. A Turing machine, operating with this character set, has two parts.

First, there is the machine itself. It is capable of being in any of a finite list of machine states, $q_1, q_2, \cdots, q_n, q_H$. Of these $(n+1)$ states, the last one, q_H, has a special role, as we shall see shortly. These machine states serve as the RAM: The machine will store data temporarily by the choice of the particular state in which it currently resides.

Second, there is a semi-infinite tape, divided into a succession of square boxes. Thus, at the beginning of the tape there appears the first box, followed, moving along the tape, by the second, then the third, etc. There is no "final end" of the tape; that is, there is available as much tape and as many boxes, so arranged, as might be needed. Each of these boxes may have printed within it a single character from the set $\mathcal{C}$, or the box may be blank (having no character). We denote this blank box-state by $\emptyset$ (not to be confused with the empty string). This tape serves as the hard drive: The machine will store data here on a more permanent basis for later use in the computation.

The machine also has a read/write head, which at any one moment resides over one of the squares of the tape. Thus, the complete state of this system (machine, tape, and head), at any moment, is characterized by specifying (i) the internal state of the machine, (ii) the characters printed on the tape, and (iii) the square over which the head currently resides. For example, a typical system state might be: "machine state q_7; tape configuration 3 \$ v Z x $\emptyset$ $\emptyset$ $\emptyset$ $\cdots$; head over the fourth square." In this configuration, the head would be read the character Z and would print to the fourth square. It will turn out that the only tape configurations of interest will be those in which all squares of the tape beyond a certain one are blank.

This machine operates by going from one system state to the next according to certain rules that are set down in a table (the "program"). A typical row in this table is as follows:

Curr State	Curr Char	$\rightarrow$	New State	New Char	Move
q_7	Z		q_3	p	L

This row demands that, if the computer finds itself in machine state q_7, with the head reading character Z on the tape, then the computer is to (i) change its internal state to q_3, (ii) erase the character Z from that square on the tape and print instead character p, and (iii) move the head one square to the left (i.e., toward the beginning of the tape). Thus, this particular row in the table would send the system from the system state given in the previous paragraph to the following system state: "machine state q_3; tape configuration 3 \$ v p x $\emptyset$ $\emptyset$ $\emptyset$ $\cdots$; head over the third square." The full table for our Turing machine will consist of many such rows. In each row, the first entry must be one of the machine states $q_1, \cdots, q_n$ (but *not* the state q_H); the second entry must be a character, or the blank, $\emptyset$; the third entry must be a machine state (with q_H allowed here); the fourth entry must be a character or the blank; and the fifth entry must be either R (right) or L (left). Finally, the full table must contain one and only one such row for every possible choice of the first two entries. Thus, if there were 10 machine states (including q_H), and the character set had six elements, then the full table would have exactly 63 ($= 9 \times 7$) rows. The Turing machine now operates in the obvious way. At each stage, it looks up its current machine state and character under the head in this table. It then reads off from the table what should be its next machine state, what the head should print on that square of the tape, and what movement the head should make (just one square, either to the right or to the left). If ever the machine finds itself in state q_H, then the machine halts (stops computing). That is why we do not allow the q_H state as the first entry of any row.

The crucial features of this design are (i) that the number of internal machines states is finite, whereas the number of tape squares is infinite and (ii) what the machine will do next depends *only* on on the current machine state and the character under the head, and not on what is printed elsewhere on the tape or whether the head resides over the fourth square or the nineteenth square.

To run a Turing machine, select the input string S and print it, one character at a time, at the beginning of the tape, leaving all the other tape squares blank. Begin with the head over the first square and the machine in initial state q_1. Now let the machine run, step by step, for each step looking up in the table what to do next. If the machine, during the course of its running, never achieves the halt state, q_H, then it will continue running forever. Well, that's life. If, however, it does eventually achieve q_H and halt, then we read the output string from the tape, starting from the first square and continuing until we reach the first blank square. Note that, during the running of every Turing machine, the tape always contains but a finite number of nonblank characters (although, of course, the number of such characters may, as S runs over all possible input strings, grow without bound). The crucial feature of this operation is that the table is to be fixed, once and for all, *before* we are given the input string S.

So, a Turing machine is a sort of minicomputer—a computer reduced to its essentials. First write the program. Then, input an initial string S. The computer computes away. Either it eventually halts, presenting an output string to you, or it runs forever, never presenting anything. A problem π over a given character set is said to be *Turing-computable* if there exists a Turing machine (i.e., a choice of the number of internal states and of the table) that computes it (in particular, never halting, no matter what the input string S), as just described. To check whether you understand how a Turing machine works, try to convince yourself that you could write a Fortran emulator of Turing. Assuming you have convinced yourself, then we may conclude (from the mere existence of such an emulator) that the Turing-computable problems comprise a subset of the Fortran-computable problems.

We give just one example of a Turing-computable problem. A string S is called a *palindrome* if it reads the same backward as forward (e.g., "$K9s4\$q\$4s9K$)". Let the character set contain y and n (to make answers easier to express); and let the problem $\mathcal{S} \xrightarrow{\pi} \mathcal{S}$ be the following: $\pi(S)$ is "y" if S is a palindrome and "n" if it is not. This problem is Turing-computable. We shall not write out the full table (which would have hundreds of rows!) that demonstrates this, but rather we merely indicate how the machine would work.

The machine, in initial state q_1, reads the first entry in the string: Say it is a K. The machine then goes into a state we call p_K (whose description is "I've just read character K, and I'm now going to check to see if this is also the last character"), prints $\emptyset$, and moves one square to the right. If the head now reads anything other than $\emptyset$, the head moves another square to the right without changing anything. (That is, the table entry for "current state p_K and current character {nonblank}" requires remaining in p_K, reprinting whatever is already in that square on the tape, and then moving one square to the right.) The head thus continues moving to the right, one step at a time, until it encounters a blank square. On encountering a blank square, the machine goes into a new state we call r_K (whose description is "I'm now ready to compare the last character of the string with K), reprints $\emptyset$, and moves one square to the left. The table entry for "current state r_K, and current character anything but K" puts the machine into a new state q_n (whose description is "This string is not a palindrome. Tough luck. I'm going to go back to the beginning now, to report that fact."). (We'll return later to how this is done.) The table entry for "current state r_K, and current character K" puts the machine into a state q_2 (whose description is "So far so good. I'll go back to the beginning now and get the next character."), prints $\emptyset$, and moves one square to the left. As long as the head continues to encounter nonblank squares, it continues to move leftward back over the string. (That is, the table entry for "current state q_2, current character {nonblank}" retains the state q_2, reprints whatever is already under the head, and moves one square to the left.) However, as soon as the head meets a blank square, the machine goes back to state q_1, prints $\emptyset$, and moves one square to the right. The process now starts over (but now with a shorter string, for we have just removed from the original string S its first and last characters). That is, the machine reads the current square (yielding, say, character 9), goes into state p_9, moves to the right until it encounters a blank square, goes into state r_9, carries out a comparison of 9 with the current character, goes into either state q_n or q_s, etc. Continue in this way. If, eventually, the string is exhausted, then the machine goes into state q_y (whose description is "It is a palindrome! I cannot wait to to back to the beginning and deliver the good news"). (That is, the table entry for "current state q_1, current character blank" places the machine in state q_y, prints $\emptyset$, and moves one square to the left.)

This process eventually places the machine in either the state q_n or the state q_y. How does the reporting of the news ("y" or "n") work? We want the machine states q_n and q_y to move the head to the left, for that is where the reporting must take place. But how will our Turing machine know when it has reached the leftmost square? (There will just be blanks back there, for we have now erased the initial portions of our original string S.) One way to accomplish this is to move, initially, the entire string S up the tape a little bit, to make room for a marker at square one. Here is how to move the entire string S one square to the right. Read the first character (say, K, again). Go to state s_K (whose description is "I'm about to move a K one square to the right"), and move the head one square to the right. If, say, the next square read by the head contains the character 9 then print the K in this square, go to state s_9, and move another square to the right. (That is, these are the instructions for "current machine state s_K, current character 9.") Continue until you reach a blank square (i.e., to the end of the string S). Then print the last character (as determined by what s-state you happen to be in at the time), and go into a state that causes movement to the left until you encounter a $\emptyset$. You have now moved the entire initial string S one square to the right. In a similar way, we may move the initial string to the right a second square, and print anything (say, v) in the first square of the tape. All of this would be done *before* the program of the previous paragraph. Do this, and then run that program (on the original string S, as now displaced).

We next describe how the reporting works. The state q_y will require motion of the head to the left to continue as long as that head encounters only blank squares. But as soon as it encounters character v on the tape (i.e., as soon as it reaches the first square of the tape), it will print y and go to the halt state, q_H. In this way it is returned that the original string S was a palindrome. The state q_n has to work a little differently. It will cause the head to continue moving to the left as long as nonblank characters are encountered. But, according to q_n, as soon as the head encounters a blank, the machine goes into still another another state, q_{nn} (whose description is "OK. Now all I need is to is find that v off to my left, to whom I must report"). So, the table entry for "current state q_{nn}, current character v" is "Go to state q_H, print character n, and move one square to the right." In this way it is reported that the original string S was not a palindrome.

Well, that was exhausting. Suppose our initial character set contained m characters. Then we must introduce $3m$ machine states, for the p's, r's, and s, as well as an additional nine machine states, for the various q's, including q_H. Thus, there will be a total of $(3m+8)(m+1)$ rows in the table. Even for just 10 characters, for instance, this is a total of 418 rows! You might think it would have been easier to have the machine simply remember the string S as it passes over it the first time, and then just make a single check for palindromeness when it reaches the far end of the string. But that will not work, because the machine is allowed only a finite number of internal states, and this number must be fixed already in the original table—and there is no adjusting that number depending on the string S.

Exercise. Why do we not, in our definition of a Turing machine, fix the total number of machine states, once and for all? Because that does not work. Fix the character set $\mathcal{C}$. Prove that, for every integer n, there exists a problem (over that character set) that cannot be solved by any Turing machine with just $(n+1)$ internal states.

Exercise. Let the character set be any one that includes the 10 digits together with n. Convince yourself that you could build a Turing machine that returns "n" if input string S is not an integer and returns $S+1$ if it is an integer. Convince yourself that you could build a Turing machine that, whenever the string S is two digits separated by a single n, returns their sum and otherwise returns "n". Convince yourself that you could build a Turing machine that not only solves the problem of the last paragraph but cleans up the tape (i.e., removes everything but the answer string) before reporting.

The next step in learning this subject is to play with Turing machines to get a feeling for what they can do. Convince yourself that you could build machines (but do not actually do it!) to solve successively harder problems. Start with easy problems, such as those discussed thus far. Then try harder ones: multiplication of integers, division of integers with a remainder, deciding whether or not an integer is prime, deciding whether or not an integer is a counterexample to the Goldbach conjecture, etc. Through this process, you must eventually convince yourself of the following key fact: *There exists a Fortran emulator in Turing.* (See, e.g., Yasuhara [14] for some details.) As a consequence of this fact, the Turing-computable problems are the same as the Fortran-computable problems, and, by similar arguments, as the C-computable problems, as the Applesoft-computable problems, etc. The idea, then, is that the Turing language is the simplest one that still has sufficient richness to generate the "right" computable problems.

Here is our main definition: A problem π, over a given character set $\mathcal{C}$, is said to be *computable* if it is Turing-computable, that is, if there exists a Turing computer T that, run on any string $S \in \mathcal{S}$, always eventually halts, returning $\pi(S)$. What you have done in the previous paragraph should convince you that this is a reasonable definition. What most people do in this subject, I believe, is "talk in terms of Turing machines, but think in terms of their favorite language (whatever it happens to be)." We emphasize that, although the psychological situation here is complex, the mathematical one is not: We define a Turing machine, and, using it, we define a computable problem.

Every problem we have discussed so far is computable (including the one that sends every string to "y" if Goldbach's conjecture is true and every string to "n" if that conjecture is false). The composition of two computable problems is computable. (This fact is useful in showing that suitable changes in the input/output grammar do not affect computability.) For π and π' computable problems, the problem π'' with

$\pi''(S) = \pi(S)\ \pi'(S)$ (concatenation of strings on the right) is computable.

Many constructions involving Turing machines rest on the following fact: Every Turing machine can be represented as a string. Here is one way to do this. Consider a Turing machine T over character set $\mathcal{C}$. Let, for example, the first few rows of the table for T be the following:

Curr State	Curr Char	$\rightarrow$	New State	New Char	Move
q_7	Z		q_H	p	L
q_{11}	$\emptyset$		q_7	\$	R
q_8	2		q_8	Z	R

The first step in rewriting this T as a string is to introduce the new character set, $\mathcal{C}'$, that results from adding one additional character, say $*$, to $\mathcal{C}$. (We are assuming here that $*$ does not denote any element of the original character set $\mathcal{C}$ itself. This element $*$ will serve as a marker, as we shall see later.) The next step is to choose a string over $\mathcal{C}$ to represent each machine state for the Turing machine T. For example, we might represent states q_7, q_H, q_{11}, and q_8 by strings "$s6$", "$\$B4$", "$uU$", and "8", respectively. Then the rows of the table would be represented by a string as follows:

$$*s6*Z*\$B4*p**uU**s6*\$***8*2*8*Z***\cdots \qquad (1)$$

We have simply written the entries in the table (replacing each machine state by its string), row by row, one after another, using the $*$ element to separate the entries. The reading or writing of a blank square is indicated by placing nothing between the two separators: $**$. Movement of the head to the left is indicated by placing nothing between the separators ($**$) and movement to the right is shown by placing a $*$ between them ($***$).

Thus, each Turing machine over $\mathcal{C}$ gives rise to some string over $\mathcal{C}' = \mathcal{C} \cup \{*\}$. The machine for the palindrome problem with 10 characters, for example, results in a string of about

4,800 characters. Note that a given Turing machine can be represented by a string in many ways (for example, by choosing different strings to represent the various machine states, or by changing the order in which the rows of the table are presented).

The time has come to simplify our language a little bit. In Sect. 3, we introduced a problem π, on character set $\{0, 1, \cdots, 9, f, y, n\}$, with $\pi(S)$ equal to "f", "y", or "n" according to whether S is not an integer greater than or equal to two, is a prime integer, or is a nonprime integer. We shall now allow ourselves to describe this as "the problem of deciding whether or not an integer is prime." Thus, in this description, it is understood that (i) the character set has sufficient characters to describe the input strings of interest (i.e., here, the digits), (ii) any strings constructed from those or other characters that are not the strings of interest (e.g., here, "007"), will be suitably branded by the problem (e.g., sent to "f"), and (iii) the outcomes of interest (here, "prime" and "not prime") will be suitably encoded as strings over our character set. We can safely ignore how such details are arranged and thereby avoid an unnecessary distraction. Next, recall, from the previous paragraph, that a Turing machine over character set $\mathcal{C}$ can be represented as a string over the character set $\mathcal{C}' = \mathcal{C} \cup \{*\}$, the extra character $*$ having been introduced as a marker. Now fix any ordering for $\mathcal{C}'$, thus obtaining, as we noted earlier, an ordering (dictionary) for the strings over $\mathcal{C}'$, and thereby an assignment of an integer to each such string. Combining these two constructions, then, we assign, to each Turing machine over $\mathcal{C}$, an integer (although, of course, not every integer arises from some Turing machine). Here is a somewhat more useful assignment. Consider the first string over $\mathcal{C}'$ (in this ordering) that represents a Turing machine, and call that machine number one; then consider the second string that represents a Turing machine, and call that machine number two; then the third; etc. In this way, we assign to each Turing machine an integer, such that now each integer also represents some Turing machine. Thus, we may speak of "the nth Turing machine," implicitly invoking this numbering. Next, we may combine this construction with our correspondence between strings over (the now ordered) $\mathcal{C}$ and integers. There results an assignment, to each Turing machine over $\mathcal{C}$, of a string over this same character set such that each string now represents some Turing machine. We denote by T_S the Turing machine associated with string S. Shortly, we will want to turn a pair, such as (T, S), where T is a Turing machine over $\mathcal{C}$ and S a string over $\mathcal{C}$, into a single string over $\mathcal{C}$. We may do this, for example, in the following manner: First take the string over $\mathcal{C}' = \mathcal{C} \cup \{*\}$ that represents T (as before), then append "$*****$" (a marker, to separate the representation of T from S), and finally append the string S. In this way, we represent (T, S) as a string over $\mathcal{C}'$. But now we may convert this to an integer—or to a string over $\mathcal{C}$—using the techniques just described. If you find yourself uncomfortable with all these conventions, you might try to restore the missing material for a short while, until you get used to them.

Exercise. Convince yourself that each of the following problems is computable: (i) that of deciding whether or not a string over $\mathcal{C}'$ represents *some* Turing machine; (ii) that of deciding whether or not two strings represent the same Turing machine [where by "same" we mean "differing only in rearrangement of the machine states (preserving q_1 and q_H) and in the order in which the rows are presented in the table"]; (iii) that which sends integer S to the string for the Sth Turing machine; (iv) that which sends integer S to the string for the Sth Turing machine, eliminating repetitions (via "same"); (v) any problem π such that $\pi(S) = \emptyset$ for all but at most a finite number of strings; and (vi) the problem that assigns, to each string S, the positive integer that is the number of steps Turing machine T takes, on string S, before it halts, where T is some fixed Turing machine that does halt for every input string. Much more difficult, for example, is the problem of deciding whether two Turing machines compute the same problem (or, indeed, whether a given Turing machine T computes any problem at all, i.e., whether that machine always halts, no matter what the input string).

Chapter 6

Noncomputable Problems

It is not hard to convince yourself that every problem is computable. A problem, after all, is merely a mapping $\mathcal{S} \xrightarrow{\pi} \mathcal{S}$. So, to specify a problem, you must specify what the mapping is, that is, specify how to determine, for any string $S \in \mathcal{S}$, some string, $\pi(S)$, or specify how to compute, given any S, some $\pi(S)$. But "compute," we have come to realize, means "Turing-compute."

Although this intuitive argument may seem plausible, it is simply wrong: There do indeed exist noncomputable problems. The easiest way to prove this is by a cardinality argument. The set of all Turing machines that compute problems is countable [since it is a subset of the set of all Turing machines, which in turn can be represented as a subset of the (countable) set of strings over some character set]. But the set of all problems, as we saw in Sect. 3, is uncountably infinite. Therefore, the mapping "send machine to the problem it computes" from the former to the latter cannot be onto.

Although this proof is simple, it does not give much insight into which problems are noncomputable and which are not. Fortunately, it turns out that there is an example that is both simple and illuminating.

> The *halting problem* is that mapping $\mathcal{S} \xrightarrow{\pi} \mathcal{S}$ that sends Turing machine T and string S to "halt" if the machine T, running on input string S, eventually halts; and to "not halt" if that machine on that string continues running indefinitely without halting.

Note that the halting problem is indeed a problem, for, given machine T and string S, then T on S either halts, or it does not. You might imagine that we could build a master Turing machine, **H**, that would compute the halting problem, in the following manner: Given (T, S), where T is some Turing machine and S some string, **H** would merely simulate the action of T on S, doing what T would do, step by step, and ultimately reporting the result: "halt" or "not halt." But, unfortunately, this does not work. There is no difficulty if T, applied to S, ultimately halts. Then **H** will discover this eventually, and duly report "halt." But what if T, applied to S, never halts? In this case there will never be a moment when **H** discovers this fact, and so no moment when **H** will report "not halt".

Now comes the central result of this subject:

> **Theorem.** The halting problem is not computable.

> Proof: For any string, S, denote by T_S the Turing machine represented by that string, as described previously. Suppose, for contradiction, that there existed a Turing machine, **H**, that computed the halting problem, reporting $\mathbf{H}(T, S)$ = "halt" or $\mathbf{H}(T, S)$ = "not halt", according to whether machine T, applied to string S, halts or not. We now construct a new Turing machine, $\tilde{T}$, as follows: Given string S, $\tilde{T}$ first runs machine **H** on (T_S, S), and then proceeds as follows: If $\mathbf{H}(T_S, S)$ = "halt", then $\tilde{T}$ continues running, without ever halting; while if $\mathbf{H}(T_S, S)$ = "not halt", then $\tilde{T}$

immediately halts. [In other words, we build a Turing machine $\tilde{T}$ that, given string S, asks **H** about (T_S, S), and then does the opposite of what **H** reports.] Now, $\tilde{T}$ is a Turing machine, and so it is represented by some string: $\tilde{T} = T_{\tilde{S}}$, for some $\tilde{S}$. We now ask the following: What happens when machine $\tilde{T}$ is run on string $\tilde{S}$? Suppose, say, that it eventually halts. But this means, from the way we defined machine $\tilde{T}$, that $\mathbf{H}(T_{\tilde{S}}, \tilde{S})$ = "not halt". But this means, from the defining property of **H**, that machine $T_{\tilde{S}}$ $(= \tilde{T})$, when run on string $\tilde{S}$, does not halt. This is a contradiction. Similarly, the supposition that machine $\tilde{T}$, run on string $\tilde{S}$, does not halt leads to a contradiction. Thus, since the assumption that there exists a Turing machine **H** that computes the halting problem leads to a contradiction, we conclude that the halting problem is not computable.

This proof—essentially, a diagonal argument—is at the same time very simple and very confusing. I urge you to return to it in the coming weeks, as often as necessary, until you have mastered it. The discussion that follows is intended to give you a feeling for what the theorem means.

Note that the theorem does not assert that there is a *specific* machine T and string S such that we will be unable to decide whether that T, run on that S, halts. Indeed, we expect that, given (T, S), we could, given enough time and ingenuity, determine whether halting occurs. What the theorem does assert is that there is no single algorithm that will correctly decide halting in *every* case, that is, for *every* (T, S).

Here is a more poignant restatement of these observations. Imagine having the following job: Occasionally, a Turing machine T and string S are brought to you, and you are to determine and report to your boss whether or not that machine, applied to that string, ever halts. In some cases—for example, a machine for which q_H never appears in the third column of the table; or for which all states in the third column are q_H—your decision will take but a few minutes. In other cases—for example, that in which there is a collection of machine states (i) from which the machine cannot exit, (ii) such that q_H does not appear in the third column for any of these states, and (iii) into which the machine, by virtue of the given S, will enter—it may take take hours. In more complicated cases it might take days... or even years. As you continue working in this job, you will build a repertoire of arguments for settling this question in specific cases. And you will note that you are continually adding new, ever more clever, arguments to your collection. At some point, you may ask yourself, "Will this job ever become routine? Will I ever reach the point at which I have developed all the arguments that are needed to solve these puzzles—the point at which no further originality will be required for this job?" These questions are answered by the theorem just stated: The answers are all "No."

Suppose for a moment that we had felt inclined to include use of the additional command "Do, for $I = 1, \infty$ $\{\cdots\}$ Next" (i.e., the infinite Do-loop) in our notion of "computable." As a result, as we have noted, there would be more computable problems. However, we could still define the halting problem (now referring to Turing machines in which this additional command is allowed). But the theorem above would still hold in this case (for its proof would go through in the same way). In other words, we would conclude that, even in this stronger language, we cannot compute the halting problem for that language.

Next, suppose for a moment that we had a master Turing machine **H** that did compute the halting problem. Then, we claim, we could resolve the Goldbach conjecture. We do this as follows: Construct a Turing machine T that, applied to any string S, ignores S completely and starts searching the even integers $(4, 6, \cdots)$ looking for a Goldbach counterexample. If it finds a counterexample, it halts, announcing this result. As long as T has not yet found a counterexample, it just keeps looking. Now, all we have to do, having built this machine T, is run the master machine $\mathbf{H}$ on machine T and any string S. If the result is $\mathbf{H}(T, S) =$ "halt", then the Goldbach conjecture is false; if the result is "not halt", the conjecture is true. Note that we settle this conjecture without doing any real work: We do not have to have deep thoughts about the structure of the primes or about any other relevant mathematics. All we need to resolve the conjecture is, essentially, an understanding of what it is asking for. In a similar way, we could use $\mathbf{H}$ to resolve, again without doing any real work, many of the other open questions in mathematics. In short, a great deal of mathematics can be encoded into the question of whether certain Turing machines halt. Perhaps this observation makes the theorem seem less surprising.

One occasionally reads, in the Sunday supplement, an article suggesting that physics is dead, that we have now discovered the fundamental structure of Nature—the "theory of everything" and that all that remains is working out the details. Of course, this is a mere guess on the part of the writer: Nobody has (or can have?) any real insight into this question. But note that mathematics is very different from physics in this regard. Mathematics is not dead yet; and, we suggest, it never will be. Indeed, we have a *theorem* to the effect that new and different insights will always be required in the development of mathematics!

Exercise. Show that the following problems are not computable: (i) the problem that asks whether a given Turing machine solves *some* problem; (ii) the problem that asks whether, given a Turing machine, there is *some* string S on which it halts; (iii) the problem of deciding whether two Turing machines (both of which do solve some problem) solve the same problem. (Hint: Show that a Turing machine that computes these problems could be reconfigured to give a Turing machine that computes the halting problem.)

This paragraph is mere whimsy, and you should feel free to ignore it. I would like to suggest that the expression "X never happens" (as well as its various siblings) has no real meaning whatsoever. Rather, this expression is merely a part of a sociological convention: We have all agreed that, when we hear it, we shall nod our heads knowingly (rather than, say, rolling our eyes). Certainly, this idea is not necessary to function in our daily lives. Wolves, for example, neither use nor understand this expression, and yet they get along, in the woods, quite well. Imagine an individual who has been raised by wolves and shares *their* sociology. You wish to explain to this person that "This Turing machine, when run on this string, never halts." This individual replies, "I have no idea what you are talking about." You say, "Well, the machine doesn't halt after nine steps." "Right." "It doesn't halt after 137 steps." "Right." "And, in fact, it doesn't halt after any number of steps." "I have no idea what you are talking about." Or, you might try to argue using the structure of a particular Turing machine T. "The state q_H nowhere appears as the third entry in any row." "Right." "Therefore, the machine doesn't halt after 19 steps, because it couldn't be in the state q_H then." "Right." "Similarly, it doesn't halt after n steps, for any $n = 1, 2, \cdots$." "I have no idea what you are talking about." Your growing sense of frustration arises from your inability to express this idea in terms of anything else.

Here is another perspective. Imagine that you were transported to another planet, the residents of which have decided to explain to you a term, "swerm," in their language. You find yourself on the other side of the conversation. They say "Horses are brown." "Right." "And three is an integer." "Right." "And swerm." "I have no idea what you are talking about." They try to argue in more detail—exploring the light reflection from horse's coats and counting up to three—but still you cannot understand. They feel a growing frustration with your skepticism. The residents of this planet, it turns out, have introduced the notion of a Turing machine, which they may use to compute problems. Particularly famous is the *swerm problem*: Given string S let $\pi(S)$ be "swerm" or "not swerm" according to whether or not the string S is swerm. You, of course, suspect that this is not a problem at all, but just nonsense talk. In fact, they have even proven what they regard as an important theorem: The swerm problem is not computable. "At last!," you think, "Surely now, by merely going through their proof, I will be able to understand what this 'swerm' is all about." So, you go to their library, find the paper containing this proof, and begin to read. But you quickly discover, to your dismay, that their so-called "proof" tells you nothing whatever—for it uses, in an essential way, the very concept of swerm. (Apparently, the people of this planet have decided that it is appropriate to use swerm within proofs.) You conclude that, at worst, the residents of this planet are delusional. At best, they have managed to discover that it is not possible to construct a Turing machine that will explicate their strange sociology.

Chapter 7

Noncomputable Numbers

As an example of an application of Turing machines, we now consider briefly the subject of noncomputable numbers.

A positive real number x is said to be *computable* if there exists a Turing machine that, when applied to any positive integer S as input, returns some rational number, a/b, such that $|x - a/b| \leq 1/S$. In other words, the computable numbers are those to which we may compute approximations. Note that the two integers a and b in the fraction must be encoded into a single string in the output (e.g., by using a separator, and then translating back to the original character set). The reason that we approximate x by rationals is that it is easy to express a rational number in terms of a string. Note also that many Turing machines may compute the same number x (e.g., by providing different rational approximations to it). And, finally, note that the function $1/S$ on the right of this equation could as well be replaced by any (computable) function of S that decreases monotonically to zero (e.g., $1/S^7$ or e^{-S}), resulting in the same notion of computable number: You can easily retrofit a Turing machine designed for one function on the right to one designed for another. The problem of whether a given Turing machine "computes" some real number x in this sense, as well as that of deciding whether two machines compute the same number, is not computable.

> **Exercise.** Call a number x *hypercomputable* if there exists a Turing machine that, given integer S as input, returns a Turing machine that computes (in the sense used here) some real number y such that $|x-y| \leq 1/S$. Clearly, every computable number is hypercomputable. Is every hypercomputable number computable?

The number e, for example, is computable. An appropriate Turing machine might use the formula $e = 1+1/1!+1/2!+\cdots$, keeping enough terms and computing the terms with sufficient accuracy to determine a rational within $1/S$ of e. Similarly, the number $\log[\sin^{-1}(.714) + \sinh(e/4)]/\pi^{2.7}$ is computable, as is every other other number you might think of offhand. Note that whether or not a number is computable depends only on the number itself, and not how that number is expressed. Thus, every rational number is computable, as is the number that is 1 if Goldbach's conjecture is true, and 0 if it is false. Indeed, it is tempting to imagine that every number might be computable. However, there do indeed exist noncomputable numbers, as follows immediately by a cardinality argument: The set of real numbers is uncountably infinite, whereas the set of Turing machines is only countable infinite.

Again, as with the case of noncomputable problems, we would like, not merely an existence argument, but a "concrete" example. Here is one. Set

$$c = \sum_{n=1}^{\infty} a_n/3^n, \tag{2}$$

where $a_n = 2$ if, for the nth pair (T, S), the result of running the Turing machine T on the string S halts, and $a_n = 0$ if that machine on that string does not halt. Note that, since each Turing machine T on each string S either halts or does not halt, each a_n is a definite number, and so this c is also a perfectly definite number. If you know this c to sufficient accuracy, then you know whether each of the first n Turing machines/strings halts. Indeed, either $c < 1/3$ (the case in which the first Turing machine/string does not halt) or $c > 2/3$ (the case in which it does halt); so knowing c within $(1/6)$ determines whether or not the first machine/string halts. Similarly, knowing c to within $1/(2 * 3^n)$ determines whether each of the first n machine/strings halts. It follows from these remarks that the number c is not computable. Indeed, suppose we were given Turing machine, $\tilde{T}$, that computes c, in the sense described here. Then, we could easily rebuild that machine into one that computes the halting problem, as follows: If you wish to know whether the nth pair (T, S) halts, apply this $\tilde{T}$ to string $2 * 3^n$ (written out as its digits), and interpret the rational number that results. But we know that the halting problem is not computable, and so no such machine $\tilde{T}$ exists. Here is a curious corollary of these observations: The number c of Eqn. (2) is not rational. Note that this is not at all obvious from that equation.

> **Exercise.** Show that there exists a Turing machine that accepts as input a positive integer S, returning a rational a/b, such that the resulting sequence of rational numbers increases monotonically and converges to c from below. [Hint: Given n, run, for each $k < n$, the kth (T, S) pair for $(n-k)$ steps.] Prove that if number x is such that there exists a Turing-generated monotonic sequence of rationals converging to it from below (in the sense of the previous sentence) and also one from above, then x is computable. Does there exist a noncomputable number such that there exists neither such a sequence from above nor one from below? What about the number that results from (2) by replacing 3 by -3 on the right?

It is interesting to speculate what might happen if ever a physical theory were to predict, for the outcome of some experiment, a noncomputable number (e.g., the c of Eqn. (2)). Then, since c really is a number, the theory would be making a perfectly definite prediction for the outcome of the experiment. However, to evaluate that predicted number, to higher and higher precision, would require new and ever more sophisticated insights (for that is the meaning of not being computable). Thus, we might some day reach the situation in which the experimentalists, who have carried out the experiment to, say, one part in 10^7, are way ahead of the theoreticians, who have only been able to carry out the computation of what the theory predicts to one part in 10^2! And there would be no guarantee that any greater precision would be forthcoming from the theoreticians any time soon. (A more accurate determination of the prediction of the theory might require, for example, that Goldbach's conjecture be settled.) This speculation is not entirely idle, for there are some (very weak) indications that noncomputable numbers may actually arise in some future quantum theory of gravity.

Chapter 8

Formal Mathematics

The most famous application of computability is to a certain program for formalizing mathematics. We here merely touch on a few highlights of this subject. For more details, see, for example, Kelly [8]. Nothing in this section will be used later, so it may be skipped.

The idea is to apply the notions of the previous sections—strings, problems, computability, etc.—to mathematics itself. But this program immediately runs into a serious roadblock: The ingredients of ordinary mathematics—the definitions, theorems, and proofs—are informal in character. Here is an example.

> **Theorem.** There is no largest prime integer.
>
> Proof: Let, for contradiction, n be the largest prime integer. Set $a = n! + 1$. Factor this integer a as a product primes, and let p denote one of those prime factors. Then p cannot be 2, for this formula shows that a is an even integer (namely, $n!$) plus 1, that is, that the division of a by 2 leaves a remainder of 1. And, similarly, p cannot be 3, for the division of a by 3 also leaves a remainder of 1, and so on, up to n. We conclude that $p > n$.

> Thus, this p is a prime number greater than n, which contradicts our choice of n.

What appears here is merely seven English sentences, so designed to cause the reader to nod in agreement. True, it is also a string (over some character set), but these are not the sorts of strings that can easily be manipulated by Turing machines, or about which we can easily prove theorems.

The idea of formal mathematics is to introduce a certain class of particularly simple strings, together with certain manipulations of those strings, that will reflect the content of ordinary mathematics. Thus, for example, certain strings will be deemed "assertions." Of course, these strings will be merely meaningless lists of symbols: They will not actually "assert" anything. And, similarly, other strings will be designated "proofs of assertions" (although they will not actually "prove" anything). This framework opens up the possibility of applying mathematics to mathematics itself, that is, of proving (informal) theorems about the assertion strings and proof strings. We further arrange matters so that the problem of deciding whether a given string is an assertion string or whether it is a proof string is computable. Thus, ultimately, we will be able to apply Turing machines and the ideas of computability to mathematics itself.

Mathematics is, essentially, set theory, and thus our goal is to formalize (i.e., render as strings) the subject of set theory. We emphasize again that mere string manipulation should not be confused with mathematical "Truth." Think of "formal mathematics" as just another area of mathematics, analogous, for example, to group theory. But instead of manipulating group elements according to the rules laid down for group theory, we shall manipulate various mathematics strings according to some other set of rules we shall lay down. Adopting this perspective is more easily said than done.

Fix a character set $\mathcal{C}$ (e.g., the set of lowercase Latin letters). We next introduce a new character set $\tilde{\mathcal{C}}$, consisting of the characters in $\mathcal{C}$, together with the following 10 additional characters: $=$, $\in$, $\neg$, $\wedge$, $\forall$, $\}$, $\{$, $:$, $)$, and $($. Next, we introduce a certain collection of strings over $\tilde{\mathcal{C}}$, called the *formulae.* The rules are the following: (i) For x and y any nonempty strings over $\mathcal{C}$, each of "$x = y$" and "$x \in y$" is a formula. (ii) For $\mathcal{A}$ and $\mathcal{B}$ any formulae, each of "$\neg\mathcal{A}$" and "$(\mathcal{A} \wedge \mathcal{B})$" is a formula. (iii) For $\mathcal{A}$ any formula, and x any nonempty $\mathcal{C}$-string, "$\forall x(\mathcal{A})$" is a formula. (iv) The two expressions in item (i) also result in formulae if either or both of x and y is instead replaced by a $\tilde{\mathcal{C}}$-string of the form "$\{z : \mathcal{A}\}$", where z is any nonempty $\mathcal{C}$-string and $\mathcal{A}$ is any formula. Using these rules, we may generate an enormous number of formulae [e.g., "$\forall x((y \in x \wedge \neg\forall s(z = y)) \wedge z \in \{w : x \in w\})$]". A crucial fact about this construction is this: The problem of deciding whether or not a $\tilde{C}$-string is a formula is computable.

The nonempty strings over $\mathcal{C}$ are called *classes* (which we think of as "sets," the name having been changed for certain technical reasons). We also give these new symbols suggestive names: "$=$" is called "equals"; "$\in$" is called "is an element of"; "$\neg$" is called "not"; "$\wedge$" is called "and"; "$\forall$" is called "for all"; and "$\{z : \mathcal{A}\}$" is called "the collection of all sets z such that $\mathcal{A}$." The purpose of these names is merely to make the strings easier to remember and to think about: These names are not to be construed as bestowing any actual meaning.

A *definition* is merely a shorthand way of a writing certain, commonly occurring, $\tilde{\mathcal{C}}$-strings. Here are a few examples of useful definitions (and their names): "$\mathcal{A} \vee \mathcal{B}$" stands for "$\neg(\neg\mathcal{A} \wedge \neg\mathcal{B})$" ("or"); "$\mathcal{A} \Rightarrow \mathcal{B}$" stands for "$\neg\mathcal{A} \vee \mathcal{B}$" ("implies"); "$\exists x(\mathcal{A})$" stands for "$\neg\forall x(\neg\mathcal{A})$" ("there exists an x such that"); "$x \cup y$" stands for "$\{z : z \in x \vee z \in y\}$" ("union"); "$x \subset y$" stands for "$\forall z(z \in x \Rightarrow z \in y)$" ("subset"); "$\emptyset$" stands for "$\{z : \neg z = z\}$" ("empty set"); and "$\{x\}$" stands for "$\{z : z = x\}$" ("a set whose only element is x"). The integers are now defined as follows: 0 is defined as $\emptyset$, 1 as $0 \cup \{0\}$, 2 as $1 \cup \{1\}$, etc. Thus, for example, 5 denotes the set with precisely the following five elements: $0, 1, 2, 3$, and 4. There is also a definition (which we shall not give) of a set ω that deserves to be called the *integers.* Note that we cannot, for example, merely write "$\omega = \{0, 1, 2, \cdots\}$," for neither "," nor "$\cdots$" are allowed symbols. We emphasize that these various definitions add nothing whatever to the logical structure (nor are we incorporating their symbols into our character set): Their only role is to make it easier to write certain long strings.

You probably once learned a version of set theory in which one begins with some basic universal set, for example, "the set of all dogs," then introduces various subsets of this basic set, for example, "the set of all brown dogs," and finally introduces the various set relations—subsets, unions, etc.—on these subsets. Here, things are structured a little differently. Think of the class x (or any nonempty C-string) as having elements (just like a set), i.e., we might have "$y \in x$". But, on the other hand, this x might also be a member of some z, i.e., we might have "$x \in z$". There is no basic "universal set," fixed at the beginning, from which flows all the other sets. All we have, instead, is this abstract hierarchy, running indefinitely in both directions: Classes are elements of classes that are themselves elements of classes, etc.; and classes have elements that are themselves classes that have elements, etc.

You must now convince yourself that every assertion you have ever made—or are likely to make—in mathematics can be translated into a corresponding formula, as defined here. This exercise is similar to that of convincing yourself that everything you would have called a "procedure" or "algorithm" can be translated into a corresponding Turing machine.

The following remarks are intended to get you started in this process. A rational number is defined, as usual, as a certain ordered pair of integers; and we may introduce the set of all rationals. A real number is defined as a certain set of rationals (namely, all rationals less than that number); and we may introduce the set of all reals. Thus, for example, the English sentence "x is a real number" is translated into a certain formula (i.e., into a certain string over $\tilde{C}$). The arithmetic operations on real numbers are defined as corresponding manipulations of these sets. A mapping from set x

to set y is defined as a set of ordered pairs, (a, b), with $a \in x$ and $b \in y$, such that every element of x is included once and only once as the first entry of one of these pairs. We now can introduce, for example, the real (or complex) functions of one (or more) real variables. Continuity and smoothness of such functions ("for every positive number ϵ there exists a number δ such that ...") is translated directly into the language of $\tilde{C}$-strings. Thus, for example, the English sentence "f is a continuous real-valued function of one real variable" is translated into a certain formula. Now consider some mathematical assertion, for example, "Every smooth vector field on the 2-sphere vanishes somewhere." This assertion will be translated into a certain formula. Within this formula will be the definition of a 2-sphere [ordered triples (x, y, z) of real numbers satisfying $x^2 + y^2 + z^2 = 1$] and of a vector field (a certain map sending each point of this 2-sphere to a tangent vector at that point). This formula will further include (using the symbol $\wedge$) the condition that this vector field be smooth. Then, after all this, there will appear in our formula "$\Rightarrow$" (actually, the $\tilde{C}$-string this represents). And then, finally, there will appear the translation of "there exists a point of this 2-sphere at which that vector field vanishes." I urge you to play around with this and other examples until you have convinced yourself that the formulae (i.e., the $\tilde{C}$) strings specified above, are rich enough to encompass the language of mathematical objects and of mathematical assertions about those objects.

The next step is to isolate a certain collection of formula, called the *axioms*. We shall not attempt to write out any axiom system (of which there are several) in detail, but rather we merely indicate what those systems look like. Typical axioms might include "$\neg\neg\mathcal{A} \Rightarrow \mathcal{A}$" and "$\mathcal{A} \wedge \mathcal{B} \Rightarrow \mathcal{A}$" (logical axioms), "$x = y \Rightarrow \forall z(z \in x \Rightarrow z \in y)$" and "$\forall z (z \in \{z : \mathcal{A}\} \Rightarrow \mathcal{A})$" (tying "$=$" and "$\{z : \cdots\}$ in with "$\in$"), "$\exists y((x \in y \Rightarrow \exists w(x \in w \wedge w \in z)) \wedge (\exists w(x \in w \wedge w \in z) \Rightarrow x \in y))$" (existence of infinite unions), "$\forall x((\neg x = \emptyset) \Rightarrow (\exists y(y \in x \wedge x \cap y = \emptyset)))$" (which will, among other things, guarantee that no class is an element of itself), and "$\exists x(\exists y (\emptyset \in x \wedge \forall z(z \in x \Rightarrow z \cup \{z\} \in x) \wedge x \in y))$" (which will, essentially, guarantee the existence of "infinite sets"). Other candidates for axioms might include formulae that reflect the axiom of choice, the axiom of the excluded middle, the axiom that every subset of $[0, 1]$ is measurable, etc. The crucial thing about these axiom systems is that they are so constructed that the problem of deciding whether or not a formula is an axiom is computable.

So, fix a suitable system of axioms. A *pruf* is a finite list of formulae, each of which is either (i) an axiom or (ii) a formula $\mathcal{A}$, such that both $\mathcal{B}$ and "$\mathcal{B} \Rightarrow \mathcal{A}$", for some formula $\mathcal{B}$, appear earlier in that list. A formula is a *thurem* if it is the last formula of some pruf.

Note that the prufs and thurems are both merely meaningless strings of symbols constructed in a certain way. They are not to be confused with the proofs the theorems of (informal) mathematics (which we think of as saying that "something is true"). You must now convince yourself that every argument you would accept as a proof in ordinary (informal) mathematics can be translated into a pruf, as defined above; and that every assertion you would accept as a theorem can be translated into a thurem. This is comparatively easy, once you have accepted that mathematical assertions can be translated into formulae, for the prufs and their thurems are structured just like the proofs and theorems of ordinary mathematics.

This, then, is the outline of a scheme for formalizing mathematics.

Now let there be given some axiom system. Then there exists a Turing machine that will decide whether a list of formulae is a pruf or not (since that machine can check whether that list satisfies the conditions for a pruf). Hence, there is a machine that, given any integer S as input, will write out a thurem and is such that every thurem is included in this list. (The machine simply tries lists of strings over $\tilde{\mathcal{C}}$ one at a time, checking for, and then reporting, those that are prufs.)

That is, we can "mechanically generate all thurems." However, there is no obvious way to check, mechanically, whether a given formula is a thurem, for, although we can certainly write a Turing machine that looks for prufs of that formula, we have no way to determine whether or not that machine will halt.

The Godel incompleteness theorem states that, for every such axiom system, one of two things is true. First, the system could be *inconsistent.* This means that there is some formula $\mathcal{A}$ such that both "$\mathcal{A}$" and "$\neg\mathcal{A}$" are thurems. Whenever this occurs, then (at least, for every reasonable axiom system) every formula becomes a thurem. Second, the system could be *incomplete.* This means that there is some formula $\mathcal{A}$ that is closed (i.e., is such that every free variable is subject to a $\forall$) and is such that neither "$\mathcal{A}$" nor "$\neg\mathcal{A}$" is a thurem. In informal terms, there is an assertion that is neither provable nor disprovable via the axioms. When this occurs, we could, of course, always add one of these to get a new axiom system—but then the incompleteness theorem again guarantees inconsistency or incompleteness of that new system. The proof of the incompleteness theorem is like the proof that the halting problem is not computable. The crucial step is that the statements "there exists a pruf of $\mathcal{A}$" and "there does not exist a pruf of $\mathcal{A}$" can, using the set ω of integers, be reflected as formulae in the formal system (just as the crucial step in the halting problem is that Turing machines can query Turing machines).

As one example, let us take as our axioms a standard system, but without the axiom of choice (the formula that represents the assertion that, for every set x of nonempty disjoint sets, there exists a set having exactly one element in common with each element of x). Then it is known that neither the axiom of choice nor its negation is a thurem of that system. We may add the axiom of choice (or its negation!) as a new axiom.

Chapter 9

Difficulty Functions

So far, we have been interested largely in which problems can be computed and which cannot. We now turn to a somewhat different set of issues, involving what resources are required for the computation process. These "resources" can be of several types—for example, of memory space, of program length, or of time. We shall be interested in the last of these, for the benefit of utilizing quantum mechanics during the computation process appears to lie in the time required for that computation. It is entirely possible that there might be other benefits.

Let us begin with a simple example. Consider a (regular) Turing machine T, which computes some problem π. Then for any string S, T, when run with S as the initial string, will eventually halt. Denote by $f(S)$ the total number of steps the machine T will execute before halting—a measure of the "time" required for the computation. We call this f the *step-difficulty function* of T. This function f clearly depends on the problem π itself, but it may also depend on the particular algorithm we implemented (via T) in the computation of π. Note that every step-difficulty function satisfies $f(S) \geq 1$ for every string S.

With this example in mind, we now introduce the following definition: Fix a character set $\mathcal{C}$. By a *difficulty function* over $\mathcal{C}$, we mean a function $\mathcal{S} \xrightarrow{f} R$, from the $\mathcal{C}$-strings to the reals, that is positive and bounded away from zero; that is, that for some number $b > 0$ satisfies $f(S) \geq b$ for every string S. Think of the number b as the time required to boot the computer: We do not wish to address the possibility that, for a couple of very simple input strings, the computer might be able to provide an answer in "zero time" or in an arbitrarily small time. The step-difficulty function of a Turing machine that solves a problem is, of course, just one example of a difficulty function. [Note that, although the step-difficulty functions are all integer-valued, we allow (for later convenience) our difficulty functions to be real-valued.]

This is of course merely a definition within mathematics, but it is our intention to apply it to certain computations—both Turing and otherwise. In light of this intended application, we realize that this definition has an unfortunate feature: The difficulty functions provide too much detail. For example, it might be argued that a Turing machine should be allotted less time for a step in which the character under the head remains unchanged than for a step in which the machine has to print a whole new character. Or, we might purchase for our Turing machine a new chip, which runs twice as quickly as the old (but, say, takes longer to boot). These changes in the computing setup would, arguably, require a different choice of difficulty function. However, although such technological improvements can certainly be important, they are not the subject of interest here. We, rather, are concerned with issues such as comparing, with respect to their difficulty, several problems or several algorithms for computing the same problem. These ideas motivate the following definition: Given two difficulty functions f and f', we write $f \sim f'$ provided that, for some number $a > 0$, $f(S) \leq af'(S)$ and $f'(S) \leq af(S)$ for all strings S. We note that this is indeed an equivalence relation on difficulty functions. It is the equivalence classes that reflect the sense of difficulty that we are concerned with here; and we shall always be interested in difficulty functions *only* up to this equivalence.

Exercise. (i) Fix a Turing machine, with step-difficulty function f, that solves a problem. Let f' be the difficulty function that results if the charge is only half a unit for a Turing step that leaves the character on the tape unchanged but still a full unit for a Turing step that prints a new character. Prove that $f \sim f'$. A similar result holds for new allocations of units depending on the machine internal state, on whether the head is to be moved to the left or right, etc. (ii) Prove that, for a any positive number, $f \sim af$ and $f \sim f + a$. (iii) Prove that, for $a < \mathrm{glb}(f)$, $f \sim f - a$ (where glb denotes the greatest lower bound). (iv) Prove that, if f and f' are equal for all but a finite number of strings S, then $f \sim f'$. (v) Characterize the functions h with the following property: Whenever $f \sim f'$, then $h(f) \sim h(f')$. (vi) Let Turing machines T and T' compute problems π and π', respectively. Then we have seen how to build from these two a new machine, T'', that computes $\pi'' = \pi \circ \pi'$. Show that the corresponding difficulties (up to equivalence) are related by $f''(S) = f'(S) + f(\pi'(S))$.

These examples show, among other things, that the equivalence classes have some very desirable properties: The difficulty equivalence class does not depend on how units are allocated for various types of Turing steps, on how much time is required for booting, on the purchase of a better chip, or on the act of learning how to treat a few S's very quickly.

We next introduce two notions that compare difficulty functions.

Let f and f' be two difficulty functions. We write $f \leq f'$ provided that, for some number $a > 0$, we have $f(S) \leq af'(S)$ for every string S. We note that (i) replacing f and f' by equivalent difficulty functions does not change this relationship; (ii) both $f \leq f'$ and $f' \leq f$ hold if and only if $f \sim f'$;

(iii) $f \leq f' \leq f''$ implies $f \leq f''$; and (iv) for f bounded above, we have $f \leq f'$ for every f'. That is, "$\leq$" has the properties one would associate with "less than or equal to." But note that, given two difficulty functions f and f', it is not necessarily the case that either $f \leq f'$ or $f' \leq f$. For example, on the positive integers, let $f(n) = \sqrt{n}$ and $f'(n) = 1+n \sin^2(n/20)$.

There is, in addition to "$\leq$," a second type of inequality on difficulty functions. For f and f' two difficulty functions we write $f \ll f'$ provided that, for every number $a > 0$, we have $f(S) \leq af'(S)$ for all but at most a finite number of strings S. We note that (i) replacing f and f' by equivalent difficulty functions does not change this relationship; (ii) $f \ll f'$ and $f' \ll f$ cannot both hold; (iii) $f \ll f' \ll f''$ implies $f \ll f''$; (iv) $f \ll f'$ implies $f \leq f'$; and (v) either of $f \leq f' \ll f''$ or $f \ll f' \leq f''$ implies $f \ll f''$. Again, these are precisely the properties suggested by the notation. Since these special meanings of "$\leq$" and "$\ll$" relate only *functions*, there will be no confusion with the usual meanings of these symbols, which relate only *numbers*.

We think of $f \leq f'$, with $f \not\sim f'$, as meaning that "on every string, f reflects no more difficulty than does f'; and there is an infinite number of strings on which f reflects strictly less difficulty." We think of $f \ll f'$ as meaning that "f reflects less difficulty than f' on every string." The following example will illustrate these ideas.

> **Example.** Consider the palindrome problem of Sect. 5. Denote by f the step-difficulty function (counting steps) for the Turing machine T described in that section. Set $L = \text{length}(S) + 1$, another difficulty function on $\mathcal{S}$. (The "+1" in this formula merely allows us to avoid treating the empty string separately.) Then we have $L \leq f \leq L^2$. The first follows because T must in any case traverse the entire string S (in order to examine the last character), and that traversal

already requires L steps. The second follows because in the worst case, when S actually is a palindrome, T must go back and forth across the string (or a substantial portion thereof) a total of L times, together with a few extra steps at the ends. Note that these relations are not approximations: They hold *exactly*. Although $L \ll L^2$, we have neither $L \ll f$ nor $f \ll L^2$. Here is another Turing machine, $\tilde{T}$, for computing this problem. Machine $\tilde{T}$ works the same as T, except that, on the first pass, it makes an extra check to see if the string S is of the form "$aaa \cdots a$". If it finds that form, then $\tilde{T}$ immediately returns to the beginning and reports "yes". Denote by $\tilde{f}$ the step-difficulty function of $\tilde{T}$. Then, for every string S that is not all a's, $\tilde{f}$ requires more steps than f (since T does not have to carry out those extra checks that $\tilde{T}$ does), but $\tilde{f}(S)$ and $f(S)$ differ at most by some numerical multiple of L (since this checking for a's requires just a few extra steps for each character in S). However, for a string S that *is* all a's, $f(S)$ is the order of L^2 (since T will have to go through the laborious process of checking for palindrome-ness), whereas $\tilde{f}(S)$ is the order of L (since $\tilde{T}$ will recognize this special form on the first pass). Note that there is an infinite number of such strings. It follows from all this that $\tilde{f} \leq f$, but neither $\tilde{f} \ll f$ nor $\tilde{f} \sim f$. We thus think of the computation represented by machine $\tilde{T}$ as "definitely (but only slightly) more efficient" than that of T. It seems plausible, intuitively, that given any Turing machine T' (step difficulty f') that computes this problem, there exists a Turing machine T'' (of step difficulty f'') that also computes this problem, such that $f'' \leq f'$ and $f'' \not\sim f'$. This means that, no matter how efficient you feel your present Turing machine is, there

> always exists one that is a little more efficient. It would be interesting to find a proof. On the other hand it is known [6] that there exists no Turing machine T' (of step difficulty f') that computes the palindrome problem and has $f' \ll L^2$. In other words, there is no way to compute the palindrome problem with efficiency substantially greater than L^2.

This example illustrates the idea that this equivalence relation and these inequality relations on difficulty functions are the "right" notions: They allow us to express, in a simple way, what we want to say, and they do not draw us into a discussion of what we do not want to say.

One could imagine inventing other, inequality-like, relations on difficulty functions. For example, one could compare averages of the values of the functions over certain strings, or consider the relative frequencies of the S's for which $f(S) \leq f'(S)$ or $f'(S) \leq f(S)$ occur. But these relations tend not to be very interesting, probably because they typically require some choice of an ordering for the strings, or they are are too sensitive to relatively benign relabelings of the strings.

Chapter 10

Difficult Problems—Best Algorithms

In this section, we discuss two results of Blum [3]. Both of these results are insensitive to the particular difficulty measure—or even language—employed; and both are proved by diagonal arguments. For ease of exposition, we shall discuss both results for the Turing case (i.e., the "machines" will be Turing machines, and the difficulty measures will be step-difficulty). But it should be noted that these restrictions are not necessary.

A "difficult" problem is, intuitively, one that requires many steps for its computation. It is easy to think of problems that appear, offhand, to be quite difficult in this sense, for example, of that which sends any integer S to the integer that is the third-to-last digit of the $(10^{S!})!$th prime. However, it is hard to be certain that this problem really is as difficult as it appears: There might, for example, be some marvelous theorem that asserts that this particular problem π merely returns "7" when S is even and "1" when S is odd. If this, or something like it, should turn out to be the case, then this problem π would turn out to be easy to compute.

Can we give an example of a problem π that is computable and is such that we can *guarantee* that any Turing machine that computes it has step-difficulty, say, $\geq (10^{S!})!$? The answer is yes, but for a silly reason. Let π be the problem that, applied to positive integer S, returns the string $a \cdots a$, where the total number of a's is $(10^{S!})!$. Then certainly the step-difficulty f of any Turing machine that computes this π satisfies $(10^{S!})! \leq f$, since it takes this many steps for the Turing machine merely to print out (never mind compute) the answer. This is not exactly what we had in mind. So, to avoid this sort of foolishness, we introduce the following definition: A problem π will be said to be *bounded* if the lengths of the strings $\pi(S)$, as S ranges through all input strings, are bounded above.

So, are there very difficult—perhaps even "arbitrarily difficult"—bounded problems? We formulate this question precisely as follows:

Assertion. Let $\tilde{f}$ be any difficulty function. Then there exists a bounded, computable problem π with the following property: Every Turing machine T (step-difficulty f) that computes π has $f \geq \tilde{f}$.

This assertion states, in other words, that you tell me how hard ($\tilde{f}$) you want the problem to be, and I will find a problem (π) such that every method of computing it (T) is at least $\tilde{f}$-difficult (i.e., has $f \geq \tilde{f}$). Unfortunately, this assertion turns out to be false.

Here is a counterexample. First note that, for any Turing machine T that computes a problem, the step-difficulty function f of that machine is computable. Indeed, consider the Turing machine T' that, on any string S, merely simulates the action of T on S, counting the number of steps T runs before halting, and reporting that number. This T' computes f. Now let $f_1, f_2, \cdots$ be a list of all computable, integer-valued difficulty functions (noting that the collection of such functions is countable, since the collection of all Turing machines is already countable); and let $S_1, S_2, \cdots$ be a list of all strings.

Now define a new function, $\tilde{f}$, on strings by

$$\tilde{f}(S_n) = n \times \max[f_1(S_n), f_2(S_n), \cdots, f_n(S_n)]. \qquad (3)$$

(In other words, the value of $\tilde{f}$ on string S_n is n times the largest of the values taken by the first n of our computable functions f_i, acting on that S_n.) This $\tilde{f}$ is our counterexample. To see this, fix any positive integer m. Then, for any $n \geq m$, we have $\tilde{f}(S_n) \geq n f_m(S_n)$ [for, since $n \geq m$, f_m is included in the functions maxed-over in Eq. (3)]. But this last inequality (for all $n \geq m$), and the fact that there is only a finite number of $n < m$, imply $f_m \ll \tilde{f}$. But the f_m exhaust step-difficulties of Turing machines that compute problems, and so there can be no such difficulty function f satisfying $f \geq \tilde{f}$.

The idea of this example is to so construct $\tilde{f}$ that it "grows very quickly as the string S gets larger—so quickly that no computable, integer-valued function (and therefore certainly no Turing step-difficulty function) can keep up with it." This growth is very fast indeed, for we can think of some pretty fast growing computable f's—for example, (for S an integer) $f(S) = 2$ to the power of 2 to the power of $2 \ldots S$ times. Well, that particular f (being, as it is, computable) is child's play in the hands of the *really* fast growing $\tilde{f}$ of the theorem. This situation may seem paradoxical at first sight: How can $\tilde{f}$ grow more quickly than any computable function, when Eq. (3) appears to be a computation of $\tilde{f}$? But closer inspection reveals that we do not actually "compute" $\tilde{f}$ above, for we cannot Turing-construct a list of Turing machines, $T_1, T_2, \cdots$ that compute the original list $f_1, f_2, \cdots$ (without, that is, computing the halting problem). Yet, although we cannot Turing-construct this sequence, it certainly does exists, for the set of all Turing machines is already countable, and so therefore is the set of all computable integer-valued difficulty functions.

So, in summary, it is possible to invent absurd levels of difficulty (such as that described by the $\tilde{f}$ of the previous example): There exist no computable problems that are *that* difficult. But what about more reasonable levels of difficulty? One might think of demanding that $\tilde{f}$ be computable, for, by the remarks just given, this condition would prevent $\tilde{f}$ from growing too fast. However, this demand must be implemented with some care. First, difficulty functions are real-valued, and so "computable" does not make sense for them. (Indeed, most real numbers are not even computable.) But note that every difficulty function $\hat{f}$ is equivalent to an integer-valued one [namely, the function whose value, on each string S, is the smallest integer exceeding $\hat{f}(S)$], and, for integer-valued difficulty functions, "computable" certainly does make sense. This suggests that we demand, of the function $\tilde{f}$ of the assertion, that it be integer-valued and computable. It turns out that, with this additional condition, our assertion is true:

> **Theorem** (Blum). Let $\tilde{f}$ be any computable, integer-valued, difficulty function. Then there exists a bounded, computable problem π with the following property: Every Turing machine T (of step-difficulty f) that computes π satisfies $f \geq \tilde{f}$.

> Proof: Let $T_1, T_2, \cdots$ be a list of all Turing machines (over the given character set), and let $S_1, S_2, \cdots$ be a list of all strings. Now fix any positive integer n, and consider the following prescription (ignoring for the moment the words in braces):

> **Prescription**(n). Attempt to run each of the first n machines, $T_1, \cdots, T_n$, in this order, on the initial string S_n, for a total of $\tilde{f}(S_n)$ steps each. If none of the {uncanceled} machines, so run, halts before reaching $\tilde{f}(S_n)$ steps, set $\pi(S_n) = \emptyset$. Otherwise, denote by T_i the first {uncanceled} machine in this list that *does* halt before reaching $\tilde{f}(S_n)$

steps.[4] Then set $\pi(S_n)$ equal to a string other than T_i's output: Set $\pi(S_n)$ = "a" if T_i, on S_n, halted with output string $\emptyset$; and set $\pi(S_n) = \emptyset$ otherwise. {Finally, cancel that T_i.}

This prescription, carried out for all values of n, defines a problem π [since it prescribes what string, $\pi(S)$, is to be assigned to each string S]. We note that this π, so defined, is bounded (since its only possible output strings are $\emptyset$ and a). Furthermore, this π is computable. This follows because we can build a Turing machine that (i) produces the sequence $T_1, T_2, \cdots$ of (*all*) Turing machines and the sequence $S_1, S_2, \cdots$ of all strings, (ii) simulates the running of the first n machines, as in the prescription, (iii) finds the first {uncanceled} machine that fails to run for at least $\tilde{f}(S_n)$ steps (here, by using the fact that $\tilde{f}$ is computable!), and (iv) sets $\pi(S_n)$ accordingly.

We now reinstate the braces. We carry out this prescription, in turn, for successive values of n: $1, 2, \cdots$. Each time this prescription (for some n-value) is carried out, that machine T_i (if any) used to set $\pi(S_n)$ is now "canceled," that is, is excluded from consideration in subsequent (i.e., larger-n) applications of the prescription. Thus, the new construction is identical to the old, except that, because of this cancellation, the list of Turing machines included at each stage may be smaller than it was before. But in any case the result is again some bounded, computable problem, π (different from the old π, with which we are no longer concerned).

---

[4]Note that T_i is the first machine in this ordered listing of machines that halts at all before the $\tilde{f}(S_n)$ steps and *not* that machine, among these n machines, that halts in the fewest steps.

This π is the problem whose existence is guaranteed by the theorem. To see that it has the required property, consider any Turing machine T (step-difficulty function f) that computes π. Then this T must appear somewhere in our list of machines: Say $T = T_7$. Consider the machines $T_1, \cdots, T_6$. Let n_0 be an integer such that every one of these six machines either was canceled already by the time n reached n_0 or never will be canceled for any n. [Such an n_0 exists: Indeed, each of the machines $T_1, \cdots, T_6$ either (i) is at some point (that is, for some specific n-value) canceled or (ii) is never canceled. Let n_0 be the largest of the specific n-values that occur in (i).] Now fix $n > n_0$, and apply our prescription to determine $\pi(S_n)$. Which, if any, machine is canceled during this application of the prescription? It could not be any of $T_1, \cdots, T_6$ (by definition of n_0). Therefore, T_7 is on the bubble: It will not be saved by cancellation of any of the machines before it in the list, and so it will be canceled if it halts before completing all $\tilde{f}(S_n)$ steps. But T_7 cannot be the one canceled either, for, by definition of $\pi(S_n)$, cancellation implies that T_7 on S_n differs from $\pi(S_n)$, while T_7 was assumed to compute π. We conclude from all this that T_7, on S_n, must run for at least $\tilde{f}(S_n)$ steps without halting. That is, we conclude that $f(S_n) \geq \tilde{f}(S_n)$. Since this holds for all $n > n_0$ (i.e., for all but at most a finite number of n), we conclude that $f \geq \tilde{f}$.

This is quite a proof. For each n, we stage a contest between the first n Turing machines, applying each to S_n and seeing who can go at least $\tilde{f}(S_n)$ steps without halting. We find the first machine that fails, arrange for π to be different from what *that* machine computes, remove that machine from further competition, and then repeat the contest for the next n. Since $\tilde{f}$ generally increases, the successive contests will generally get harder and harder. In this way, π avoids the losers (the machines that halt early) and thus emerges as a problem that can only be computed by a consistent winner—a machine with step-difficulty satisfying the condition of the theorem. Note that the n_0 in the proof is not computable. Note also that computability of $\tilde{f}$ is used at a critical place: to get computability of π.

> **Exercise.** Show that the previous theorem continues to hold if the last formula in its statement is replaced by $f \gg \tilde{f}$. Does there exist a Turing machine that accepts as input the Turing machine that computes $\tilde{f}$ and returns a Turing machine that computes a problem π whose existence is guaranteed by the theorem?

So, there are some pretty hard problems out there. We now turn to a related issue. It would be of great interest to define, for any given problem, a difficulty intrinsic to a problem itself (rather than to whatever method is currently being used to compute that problem). A possible line for introducing such a notion would be to let the "intrinsic difficulty" of a problem mean the minimum step-difficulty function of Turing machines that compute that problem. However, to implement such an idea we would need some result to the effect that this minimum is actually achieved. One result that would certainly do the trick is the following:

Conjecture. Let π be any computable problem. Then there exists a Turing machine T (step-difficulty f) that computes this problem, with the following property: Given any other Turing machine T' (step-difficulty f') that computes this problem, we have $f \leq f'$.

Then we would take the f of the conjecture as a measure of the intrinsic difficulty of the problem π. Unfortunately, this conjecture is false. Indeed, even for the case of the palindrome problem we have observed that, for any Turing machine T (step-difficulty f) we can think of offhand to compute this problem, there exists another machine T' (step-difficulty f') with $f' \leq f$ and not $f' \sim f$. (Embarrassingly enough, we do not understand even this simple little problem well enough to generate from it an actual counterexample to this conjecture!) Here, however, is a possible alternative conjecture—weaker than the one just given, but perhaps retaining enough strength to salvage some sort of notion of the intrinsic difficulty of a problem.

Conjecture. Let π be any computable problem. Then there exists a Turing machine T (step-difficulty f) that computes this problem, with the following property: There is no Turing machine T' (step-difficulty f') that computes this problem, such that $f' \ll f$.

This conjecture, for example, is true for the palindrome problem. Indeed, we gave a Turing machine that computes this problem with a difficulty function f with $f \leq L^2$, but not $f \ll L^2$; and it is known [6] that there is no Turing machine that computes this problem with difficulty $\ll L^2$.

Thus, this last conjecture looks promising. But, much as we might wish it to be otherwise, this conjecture is also false. Indeed, we have

> **Theorem** (Blum). There exists a computable problem π with the following property: Given any Turing machine T (step-difficulty f) that computes π, there is another Turing machine T' (step-difficulty f') that also computes π, such that $f' \ll f$.

Thus, according to this theorem, for this particular problem π, no matter how much effort you put into finding an efficient machine of computing π, there always exists a much more efficient machine waiting in the wings. You can, if you wish, submit that new, more efficient machine to the theorem, and it will then go ahead and guarantee the existence of a still more efficient machine, and so on. Thus, for the problem π of the theorem, there is an infinite succession of ever more efficient Turing machines that compute it. There would seem to be no hope of defining an "intrinsic difficulty" for this problem, at least.

We shall merely sketch the proof of the theorem. First, let h be the integer-valued function on nonnegative integers defined by $h(0) = 1$ and, for $n > 0$, $h(n) = 2^{[h(0)+\cdots+h(n-1)]}$. This function is rather rapidly growing: $h(1) = 2$, $h(2) = 8$, $h(3) = 2048$, $h(4)$ would take about 10 lines to write out, and $h(5)$ could not be written on all the paper ever manufactured on this planet. We next construct the problem π of the theorem as follows: This construction is identical to the construction of the problem π in the proof of the previous theorem, with just one small change. In carrying out the prescription, for some n-value, instead of running each of the machines $T_1, T_2, \cdots, T_n$ for the same number, $\tilde{f}(S_n)$, of steps, we now run the ith machine in this list for $h(n - i)$ steps. Thus (since h is rapidly growing), the early machines in the list are run for vastly more steps (to see if they halt) than are the later machines in the list. In any case, the result, after this one change, is a certain computable problem π (which is different, of course, from the π of the previous theorem). This π, believe it or not, has the property required in the theorem.

To see this, let Turing machine T (step-difficulty f) compute π. Then this T must be one of the T_i in our list, say $T = T_7$. By construction, using the same argument as in the previous proof, it follows that $f(S_n) \geq h(n - 7)$ for every n. We now introduce a new problem, π'. We set $\pi'(S_1) = \emptyset, \cdots, \pi'(S_7) = \emptyset$. For $n > 7$, we define $\pi'(S_n)$ by exactly the same prescription that defined π, except that we use for our list of machines, not the $T_1, \cdots, T_n$ as was used before, but rather just $T_8, T_9, \cdots, T_n$. This π' is of course also computable.

We next note that π' and π are actually equal on all but at most a finite number of strings. This follows because for sufficiently large n, say, $n \geq n_0$, each of $T_1, \cdots, T_6$ that ever will be canceled in the computation of π has already been canceled (whereas T_7, of course, will never be canceled). Once no more cancellation of these seven machines is possible, then π' and π are left to examine precisely the same machines at each step, namely, $T_8, \cdots T_n$, and so these two will end up with the same values.

We next introduce a Turing machine T' that computes π in the following manner. For $n \leq n_0$, T' simply simulates T, in this way finding out what $\pi(S_n)$ is, and returns that string. On the other hand, for $n > n_0$, T' computes π' in the manner just described (i.e., by using, in the prescription at each stage, only machines $T_8, \cdots, T_n$). Denote by f' the step-difficulty of this T'.

Finally, we claim that $f' \ll f$. It suffices to compare these two difficulty functions on S_n with $n > n_0$ (since these S_n include all but a finite number of strings). Fix $n > n_0$. Then, to compute $\pi'(S_n)$ [$= \pi(S_n)$], machine T' must simulate Turing machine T_8 (on S_n) for $h(n-8)$ steps, machine T_9 for $h(n-9)$ steps, and so on up to machine T_n for $h(0)$ steps. Thus, T' must run a total of not more than $h(0) + h(1) + \cdots + h(n - 8) = \log_2[h(n-7)]$ steps, where the last equality follows from the construction of h. Thus, we conclude that for $n > n_0$, $2^{f'(S_n)} \leq h(n-7) \leq f(S_n)$, where the last step is the bound on $f(S_n)$ found earlier. The result follows.

The first thing to notice about this argument is that it contains a flaw: Right at the end, we are comparing the number of steps that T *actually executes* with the number that T' must *simulate*. Simulating looks like a lot more work than merely executing. But for reasonable difficulty measures in reasonable languages (although not for step-difficulty in Turing) a machine can be simulated in the same number of steps (up to equivalence) as it can be run. In these cases, which include all those of serious interest, the argument is complete.

However, for the Turing case, a further, somewhat complicated work-around is necessary, which we shall not discuss.

Actually, we prove more than is stated in the theorem, namely that $2^{f'} \leq f$. In fact, one can obtain a similar result for other, specific, choices of an inequality relating f and f', by simply changing the choice of the function h. It is interesting to note that, although the theorem guarantees the *existence* of T', it does not tell us how to compute it. The crucial noncomputable step is that in which n_0 is found. In fact, there exists no Turing machine that, with input a Turing machine T that computes the π of the theorem, returns a Turing machine T', the existence of which is guaranteed by the theorem.

In summary, the prospects for assigning to each problem an "intrinsic difficulty," in some reasonable way, look pretty dismal. It may be possible to do better by some appropriate restriction on the class of problems considered. Or, there may be some way to take the greatest lower bound of the difficulty functions for machines that compute the problem, even though that lower bound is itself not realized by any machine.

Chapter 11

A Language for Efficiency

Clearly, Turing machines are highly inefficient. The key problem is that storing scratch work on a single long tape requires that the machine plod, again and again, over the same portion of tape, looking for one little piece of data after another. In the case of the palindrome problem, for example, no Turing machine can compute this problem in step-difficulty $\ll L(S)^2$; yet we might expect a "normal" computer to require only $L(S)$ steps. Thus, Turing step-difficulty functions tell us too much about Turing language and too little about the subject of real interest: the "intrinsic difficulty" of the problem or algorithm. It is time to upgrade.

We might do so, for example, to Fortran. We would assign, in some "reasonable" way, a number of steps to each Fortran command and thereby arrive at a Fortran-difficulty function for each Fortran-computed problem. We could, of course, do the same for C or other languages. Although these new difficulty functions would certainly be more realistic than Turing step-difficulty, the danger remains that they, too, would manifest excessive language dependence. But it seems, intuitively, that such dependence may be small, or—if things are set up carefully—even absent. One might imagine, for example, that we could write a C emulator in Fortran that is difficulty-function-preserving.

This situation with respect to difficulty, then, is very like that we faced earlier with respect to computability: There appears to be a universal notion lurking in the background, but that notion finds expression through many languages. We want to distill out the notion itself. The answer, in the case of computability, was Turing machines. We find the simplest language that is still rich enough to encompass our idea of computability, and then *define* computability in terms of that language. We would now like to do the same thing for difficulty. That is, we would like to invent a language, with an associated difficulty function, that is as simple as possible, but not so simple that it generates unnecessary inefficiencies. In short, we want to find a language that is to difficulty as Turing language is to computability. It will turn out, unfortunately, that our innate sense of what is the "correct" difficulty function is somewhat less firm that that of what is "computable." But, in any case, we propose, below, a language that seems to capture a more or less reasonable notion of "difficulty." There may very well be better proposals.

Fix a character set $\mathcal{C}$. For S any string over $\mathcal{C}$, we write $L(S)$ for the number of characters in the string S plus one. (The "plus one" is so we do not have to treat $S = \emptyset$ as an exception.)

Let there be created an infinite number of storage locations, each labeled by some string over $\mathcal{C}$ and each capable of holding an arbitrary string over $\mathcal{C}$. Thus, we impose no upper bound on the number of storage locations being utilized, nor on the lengths of the strings in the various locations (although each location, at any one moment, contains merely a string, i.e., a finite sequence of characters; and it will turn out that only a finite number of storage locations are in play at any one moment). We write $C(S)$ for the string in the location labeled S. The idea here is that in this way we create an ample amount of highly accessible storage space.

In the present language there will be commands, each of which directs that a certain action (mostly involving what is stored in certain locations) be taken. There are five classes of commands in this language: two for input/output, two for manipulating strings, and one for branching. These are listed below (with, for each, a brief explanation of what is to be done; and, in braces, a number representing the "difficulty" of the command, which we shall discuss shortly).

A command results if, in any of the following five items, "S" is replaced by any explicit string, "x" by any explicit character, and "n" by any (positive or negative) explicit integer:

1. INPUT TO $C(S)$: Allows the user to enter any string, which is then placed in location S. $\{L(\text{whatever string is entered})\}$

2. OUTPUT FROM $C(S)$: Allows the user to retrieve the string stored in location S. $\{L(C(S))\}$

3. APPEND x TO $C(C(S))$: Replace whatever string is stored in location $C(S)$ with that same string, but with character x appended on the right. $\{L(C(S))\}$

4. DELETE LAST OF $C(C(S))$: Replace whatever string is stored in location $C(S)$ with the string that results from deleting its rightmost character (if any). If $C(C(S)) = \emptyset$, do nothing. $\{L(C(S))\}$

5. IF (LAST $C(C(S)) == x$) SKIP n LINES: If the last character (if any) of the string in location $C(S)$ is x, then skip forward n program lines (if n is positive) or backward $|n|$ lines (if negative). If $C(S) = \emptyset$, or if the last character of $C(S)$ is other than x, or if the line to be skipped to is an INPUT command, or if there are insufficient lines in the program to carry out the indicated skip, do nothing. $\{L(C(S))\}$

By "explicit," above, we mean that the character x or the string S or the integer n must actually be written out, within the command, as some specific character or string: It cannot be indicated only implicitly, for example, as whatever happens to be stored in some location.

A *program* is a finite ordered list of commands, with the following property: The program contains exactly one INPUT command, and it is the first command of the list, and exactly one OUTPUT command, and it is the last command of the list. Here is an example of a program:

INPUT $C(abc)$
APPEND d TO $C(C(yzr574))$
IF (LAST $C(C(m)) == a$) SKIP -1 LINES
OUTPUT $C(yes)$

To run a program, place $\emptyset$ in every storage location, begin at the first program line (INPUT), and enter any string. The machine then carries out the instruction of each command in turn, then moving on to the next command in the list [except for the case of command 5 (IF), for which the next command to be executed is the one indicated above]. If and when the machine reaches the last command (OUTPUT) of the list, the machine halts, allowing the user to read the output string.

Any program, run on any input string, either halts or does not halt. If it halts for every input string, then that program *computes* some problem π, where $\pi(S)$ is the output string when string S is entered at INPUT. The program just given, for example, indeed computes a problem, namely, that with $\pi(S) = \emptyset$ for every string S. Note that we have so structured the commands that the program cannot "hang" within a single command: As long as the command follows the grammatical rules given here, then—no matter how pointless that command might be—the machine will always do *something* (or maybe nothing, as the case may be) and move on. Failure to halt can only occur by continuing to execute command after command, indefinitely, as in the following example:

INPUT $C(x)$
APPEND d TO $C(C(x))$
IF (LAST $C(C(x)) == d$) SKIP -1 LINES
OUTPUT $C(x)$

We could have modified the way the IF command works, in the following manner: We could, first, require that each command in the program be labeled by a unique string. Then, we could revise the command IF so that rather than directing that some number of program lines be skipped, it directs that there be executed next that command labeled by some explicit string. Clearly, this modification adds nothing new.

The numbers in braces, accompanying each of these five commands, give the number of "steps" we deem the computer to require to execute that command. We call this number the *difficulty* of the command. For a program that, acting on a certain string, halts, we call the total number of steps executed the *difficulty* of that run of the program; and, for a program that computes a problem, we call the total number of steps executed before halting (a function now of the input string) the *difficulty function* of the program. As always, we are interested in difficulty functions only up to equivalence. There follows a discussion of the difficulties assigned, above, to the five classes of commands.

If, in response, to an INPUT command, a string of, say, 13 characters is entered, then the execution of that command requires 14 steps. This surcharge for entering long strings turns out to be very convenient (for example, already in the following paragraph).

The number of steps assigned to the OUTPUT command is L(the string returned). This formula was chosen merely for aesthetics: Even changing it, for example, to 1 would result in equivalent difficulty functions. To see this, first note that, for any program on any string that runs up to the OUTPUT command, the total difficulty up to that point will be greater than or equal to the length of the longest string stored. (This follows since each command adds at least as much to the cumulative difficulty as it adds to the length of the longest string.) Thus, changing the difficulty for the OUTPUT command to 1 would, at most, reduce the total difficulty function by a factor of 2. But such a reduction results in an equivalent difficulty function.

For the APPEND command, we append a character to the string in the location given by the string in the location S. We have to look up location S to find $C(S)$, and then look up location $C(S)$ to find the string to be appended. Think of the difficulty, $L(C(S))$, of this command as a "lookup charge." Why is not the formula instead $L(C(S)) + L(S)$? That is, why is there no charge for "looking up" S? The reason is that this change results in an equivalent difficulty function. Indeed, in any given program there will be a finite number of APPEND commands, and so a finite number of explicit strings S in those commands. So, there will be a longest such string, say seven characters. Thus, a change in the difficulty of APPEND to $L(C(S)) + L(S)$ will add at most eight steps for this command, that is, will increase the difficulty for this command by a factor of at most 9. As a result, the final difficulty function for this program will increase by at most a factor of 9. But, such an increase results in an equivalent difficulty function. Note that the same argument does not apply to the term $L(C(S))$ in the difficulty of APPEND: This number (one more than the number of characters stored in location S) depends on what happens to be stored in S at the time, and so it cannot be bounded *a priori*. Thus, this term may make a nontrivial contribution to the final difficulty function. Why not include a term $L(C(C(S)))$ in the difficulty function of the APPEND command? After all, we have to travel to the end of the string $C(C(S))$ to append the x, and there should be some travel allowance. This is a reasonable position, which might be worth pursuing. But we have to make some decision here, and we have elected the viewpoint that there has been constructed some sort of pointer that allows us to find the end of the string easily. Similar remarks apply to the DELETE and IF commands.

In the IF command, why not include also a term $|n|$, the number of command lines skipped? After all, skipping lines is hard work, and there should be some compensation. But, again, a given program has but a finite number of IF commands, each with an explicit n, and so a maximum value of $|n|$ for all such commands in the program. Therefore, such a change in the difficulty for the IF command would always result in an equivalent difficulty function.

We now have to deal with two matters. First, we need to show that the computable problems in this new language are precisely the computable problems (defined earlier using Turing machines). Second, we would like to argue that the difficulty functions generated by this language are "reasonable," that is, that they correctly capture our intuitive sense of what the difficulty "should" be. We shall attempt to resolve both of these matters in one sweep, by generating a list of illustrative subroutines (i.e., of short program fragments). On the one hand, these subroutines will show the richness of what can be computed in this language. On the other hand, the difficulties of these subroutines, computed from the command difficulties just described, will illustrate the typical difficulty functions this language generates. In the following subroutines, $S, S', \cdots$ stand for any explicit strings, x for any explicit character, and n for any explicit integer:

1. APPEND x TO $C(S)$ $\{1\}$
2. DELETE LAST OF $C(S)$ $\{1\}$
3. IF (LAST $C(S) == x$) SKIP n LINES $\{1\}$

For subroutine 1, let, say, $S =$ "yzr". First, rewrite the program, if necessary, so that location as well as some location, say, "$h8$", are not used elsewhere in that program, so

$C(\emptyset) = \emptyset$ and $C(h8) = \emptyset$. Then APPEND y TO $C(C(h8))$; APPEND z TO $C(C(h8))$; APPEND r TO $C(C(h8))$; APPEND x TO $C(C(\emptyset))$; DELETE LAST OF $C(C(h8))$; DELETE LAST OF $C(C(h8))$; DELETE LAST OF $C(C(h8))$. The first three lines achieve $C(\emptyset) =$ "yzr"; the last three restore $C(\emptyset)$ to $\emptyset$. The total difficulty, for any one instance of this subroutine, is some fixed integer (in this example, 10), depending only on the explicit string S, and not on what is in the various memory locations at the time. But this subroutine can appear at most a finite number of times in any program, and so the actual difficulty contributed by this subroutine, each time it is run, is bounded above. So, we may assign this subroutine a difficulty 1, up to equivalence of difficulty functions. Similar remarks apply to subroutines 2 and 3.

4. SKIP n LINES $\{1\}$
5. IF $(C(S) == \emptyset)$ SKIP n LINES $\{1\}$
6. IF $(C(S) == C(S'))$ SKIP n LINES $\{L(C(S)) + L(C(S'))\}$
7. SET $C(S) = \emptyset$ $\{L(C(S))\}$

For subroutine 4, use APPEND a TO $C(\emptyset)$; IF (LAST $C(\emptyset) == a$) SKIP n LINES; and then place DELETE LAST OF $C(\emptyset)$ as the first command executed after the skip. For subroutine 5, use the commands IF (LAST $C(S) == x$) SKIP... as x runs over all possible characters, arranging the skips so that we skip n lines if all the IFs fail, but merely proceed to the next line if any succeeds. The difficulty of this subroutine, 1, results from the fact that the total number of characters is fixed. For subroutine 7, use, repeatedly, DELETE LAST FROM $C(S)$, in conjunction with subroutine 5 [to test whether $C(S)$ is empty yet]. Note that subroutine 7 has a variable difficulty: Its value depends on how many characters will have to be removed from $C(S)$.

For the following subroutines, we suppose that we begin with $C(S) = \emptyset$. If this location were not empty, then it would be necessary to use subroutine 7 first, to achieve $C(S) = \emptyset$;

and to adjust the difficulty appropriately.

8. SET $C(S) = S'$ $\{1\}$

9. SET $C(S) = C(S')$ $\{L(C(S'))\}$

For subroutine 9, we first use IF (LAST $C(S') == x$) SKIP. . . , skipping to the command APPEND x TO $C(h8)$ [where "$h8$" is some location with $C(h8) = \emptyset$]. Continue to test in this way each possible candidate, x, for the last character of $C(S')$. Then DELETE LAST OF $C(S')$, test whether $C(S') == \emptyset$ (subroutine 5), and, if not, repeat. In this way, we place $C(S')$, with its characters in reverse order, into $C(h8)$. Now do this all again, placing $C(h8)$, in reverse order, into $C(S)$. It should be clear at this point that we can carry out complicated string manipulations. For example, we can place in $C(S)$ every other character of $C(S')$, up to the first occurrence of a, and with each c replaced by $8k$, provided that $C(S'')$ contains at least six characters not including the combination yzr; otherwise. . . .

For the next three subroutines, we assume that the digits, $0, 1, \cdots, 9$, are included in the charactor set; that strings subject to arithmetic operations are already integers; and that, again, we begin with $C(S) = \emptyset$.

10. SET $C(S) = C(S') + C(S'')$ $\{L(C(S')) + L(C(S''))\}$

11. SET $C(S) = C(S') * C(S'')$ $\{L(C(S')) * L(C(S''))\}$

12. SET $C(S) = L(C(S'))$ $\{L(C(S'))\}$

For subroutine 10, for example, we first use IF (LAST $C(S') == x$) SKIP. . . and IF (LAST $C(S'') == y$) SKIP. . . , for the 100 possible combinations of digits substituted for x and y, placing, for each combination, the appropriate digit in $C(h8)$, say, as well as a marker in $C(h9)$, which tells whether or not we are carrying the 1. Then DELETE LAST OF $C(S')$, DELETE LAST OF $C(S'')$, test whether either $C(S') = \emptyset$ or

$C(S'') = \emptyset$, and repeat. We will end up with the sum, with digits in reverse order, in $C(h8)$. Now transcribe $C(h8)$ into $C(S)$, reversing the order of digits. For subroutine 11, use the usual pencil-and-paper multiplication [in the course of which each digit of $C(S')$ must be multiplied by each digit of $C(S'')$]. Using similar techniques, we can write write subroutines for loops, for example, WHILE and DO, and also for complicated branchings, such as IF ((... AND NOT ...) OR ...) CARRY OUT ...; ELSE CARRY OUT ...

It should be clear by this point that a problem is computable in this language if and only if it is (Turing) computable. After all, we have in this language the ability to enter and recover strings (INPUT, OUTPUT), the ability to manipulate strings (APPEND, DELETE) freely, and the ability to branch (IF).

Note that all of the subroutines 4–12 were constructed solely from commands 1 and 2 together with subroutines 1–3. That is, we have not so far used commands 3–5, other than to construct subroutines 1–3. Why, then, did we not omit commands 3–5, making our basic commands consist instead of commands 1 and 2, together with subroutines 1–3? The role of the "$C(C(S))$" in commands 3–5 is to allow indexed arrays (which, as we shall see shortly, play a role in efficient programming). Here are three subroutines that use this feature in an essential way:

13. SET $C(S'1)$ = FIRST CHARACTER OF $C(S)$, $C(S'2)$ = SECOND CHARACTER OF $C(S)$, ETC. $\{L(C(S)) \log(L(C(S)))\}$
14. SET $C(S) = C(C(S'))$ $\{L(C(S')) * L(C(C(S')))\}$
15. SET $C(S) = C(C(C(S')))$ $\{L(C(C(S'))) * (L(C(S')) + L(C(C(C(S')))))\}$

In subroutine 13, we are assuming that the character set contains the digits, and "$S'2$" means the string resulting from appending the character 2 to the string S', etc. Thus, this subroutine allows us to place the individual characters of the string $C(S)$ in separate locations. This makes those characters directly accessible [without having to go through all of $C(S)$ each time a character is needed]. The factor $\log(L(C(S))$ in the difficulty reflects the fact that the length of the locations ($S'n$, for $n = 1, 2, \cdots$) increases logarithmically as the length of $C(S)$. Note that the base for this logarithm is irrelevant, up to equivalence. Subroutine 15 shows that we can index arrays with indexed arrays. This subroutine is given by SET $C(e8k) = C(C(S'))$; SET $C(S) = C(C(e8k))$; and this construction yields the indicated difficulty.

> **Exercise.** Explain how to write a subroutine SKIP $C(S)$ LINES [which, say, does nothing if $C(S)$ is not an integer]. What is its difficulty?

This concludes our treatment of the present language. We conclude this section with a few additional remarks.

First note that, for any program that computes a problem, the difficulty function is $\geq L(S)$. This follows because INPUT already imposes a difficulty equal to the length of the string entered plus 1. Next, consider two programs, which compute problems π and π'. Then it is easy to write a program that computes problem $\pi \circ \pi'$: Simply juxtapose the two programs, and remove the two lines where the OUTPUT of one abuts the INPUT of the other (and, possibly, change a few explicit strings). The difficulty of the new program is the sum of the difficulties of the two components. It is easy to write short programs that change the grammar of inputs and outputs: Encoding "yes" and "no" in different ways, changing the number base, changing the character set, using character orderings in various ways, rejecting uninteresting inputs, inserting and removing separators, etc. These always have difficulty $L(S)$, where S is the string entered. It follows from all these remarks, taken together, that the difficulty function (up to equivalence) is independent of the input–output grammar.

For f and f' difficulty functions, denote by $\text{glb}(f, f')$ the function whose value, for each string S, is the smaller of the values of $f(S)$ and $f'(S)$. Then $\text{glb}(f, f')$ is also a difficulty function, and, up to equivalence, depends only on the equivalence classes of f and f'. We have $\text{glb}(f, f') \leq f$ and $\text{glb}(f, f') \leq f'$, and $\text{glb}(f, f') \sim f$ if and only if $f \leq f'$. Now let π be a problem, and let $\mathcal{P}$ (with difficulty function f) and $\mathcal{P}'$ (with difficulty function f') be programs that compute π. Then there exists a program, $\mathcal{P}''$, that computes π, with difficulty function $\text{glb}(f, f')$. This $\mathcal{P}''$ is constructed as follows: Program $\mathcal{P}''$ first makes a copy of the initial string S, then simulates the running of $\mathcal{P}$ on S for 10 steps, then the running of $\mathcal{P}'$ on the copy of $\mathcal{S}$ for 10 steps; it then continues the simulation of $\mathcal{P}$ for 10 more steps, then $\mathcal{P}'$ for 10 more steps, etc. Eventually, during these interlaced simulations, $\mathcal{P}''$ will detect a halt, and when it does so $\mathcal{P}''$ itself halts, returning the appropriate output string.

Here is a program that computes the palindrome problem. First INPUT $C(zz)$ [difficulty $L(S)$, where S is the string entered]. Then dump $C(zz)$ into $C(zzz)$, with the order of the characters reversed [difficulty $L(S)$]. Then use IF $(C(zz) == C(zzz))$ SKIP... $\{L(S)\}$ to check for palindromeness. This program has difficulty function $L(S)$. So, by the previous discussion, this program is at least as efficient as every program computing this problem. Note also that this program has difficulty function $\ll$ the step-difficulty for the Turing computation.

Here is a naive program that computes whether or not a string is prime. To compute whether integer m divides integer n (by the usual long-division method) requires $L(m)(L(n) - L(m) + 1)$ steps [for we have to multiply m by a digit ($L(m)$ steps) a total number of times given by $(L(n) - L(m) + 1)$]. So, we merely check whether the integer $n \geq 2$ entered is divisible, in turn, by each of the integers $2, 3, \cdots, \sqrt{n}$. The difficulty function of this program [at most $\sqrt{n}$ runs, each of difficulty not exceeding $(\log n)^2$] is $\leq \sqrt{n}\,(\log n)^2$ (but is not equivalent to this function, for, e.g., the even integers will be disposed of very quickly by this program). It is easy to write programs that are more efficient than this naive one, for example, by checking first to see whether n is a perfect square, and only if this fails looking for factors of n, as before. In fact, there exist programs (based on very different methods) that are *much* more efficient than that described here [1, 9].

> **Exercise.** Find a program that computes whether or not a positive integer is a perfect square; and find its difficulty function.

Conjecture. Given any program (difficulty function f) that computes whether or not an integer is prime, there exists another program that computes that problem, whose difficulty function, f', satisfies $f' \leq f$ and $f' \not\sim f$.

We remark that we could have introduced this language, instead of Turing language, right from the beginning, using it, instead of Turing, as the definition of "computable." Had we done so, then the determination of what can be computed would have been considerably simpler, if perhaps somewhat less illuminating.

Chapter 12

Are There Better Languages?

Recall that our goal is to obtain the simplest possible language that still captures what we hope is a universal notion of "difficulty." The language constructed in the previous section is intended, as we noted, as merely a suggestion. Here, we comment on a few possible alternatives.

What about dispensing with indexed arrays altogether, that is, replacing the APPEND, DELETE, and IF commands with subroutines 1–3? This would simplify everything, including the difficulty functions. But, we claim, doing so will likely result in a genuine loss of efficiency. Here is an example. Let the input S be a sequence of digits, and set $m = L(S)$. (This will be easier to follow if you think of m as being about 1,000,000, so S is written down, say, in a book of some 200 pages.) Now set, for $1 \leq x \leq m$, $f_m(x) = x^{\text{digit}(x)}+1 \bmod(m)$, where digit$(x)$ means the xth digit of S. Thus, $f_m(x)$ is also an integer between 1 and m. The problem is now the following: Let there be given some input string S. Start with $x = 7$, then find $f_m(7)$, then find $f_m[f_m(7)]$, etc, up to a total of m iterations. Report the result. Let us first compute this problem without benefit of indexed arrays. To determine $f_m(x)$, we must (i) find digit(x) (m steps, since we must search through

S) and then (ii) raise x to a small power [$\leq (\log m)^2$ steps, since x contains at most $(\log m)$ digits]. So, the difficulty to compute $f_m(x)$ is $\leq m$, and so the total difficulty to compute the problem [which entails computing $f_m(x)$ m times] is $\leq m^2$. But with indexed arrays, we may first dump the characters of S into individual locations (via subroutine 13), for a one-time difficulty of $m \log m$. But having done this, we see that computing $f_m(x)$ requires only $(\log m)^2$ steps [one log for locating digit(x) and one log for taking the power]. This yields a final difficulty function of $m(\log m)^2$. Thus, using indexed arrays is much more efficient than not. The idea of this example is that computing this problem requires that we repeatedly find characters in S, and things are so arranged that which character is to be found is almost random, making it, apparently, impossible to do all the "finding" on a single pass or two through S. It thus becomes more efficient to dump the characters of S into an array, once and for all at the beginning. The resulting easy access to the characters of S ultimately pays off. Of course, we have not *proved* that there exists no equally efficient way to compute this problem, without indexed arrays although this looks unlikely. So, the critical issue here is whether our intuitive sense is that the difficulty of this problem should be m^2 or $m(\log m)^2$. If it is the latter, then we must retain indexed arrays.

Even if we begin with commands 1 and 2 and subroutines 1–3, we could still recover indexed arrays in a simpler way: Introduce two additional basic commands, SET $C(S) = C(C(S'))$ and SET $C(C(S')) = C(S)$. These would allow us to transfer strings currently in indexed arrays to regular locations for further processing, and then to transfer the results back again to the indexed array. What difficulty shall we assign to these commands? We might use $L(C(S')) * L(C(C(S'))$, the difficulty of current subroutine 14. If we do this, then the new language will, apparently, be less efficient than the old. If, for example, we merely want to deal with the last character of a string in an array, $C(C(S'))$, then the original language permits this in just $L(C(S'))$ steps (lookup charge only), whereas the new language requires that the entire string be copied into a regular location before its last character is accessed. We could avoid this by making the difficulty, for our two new commands, just $L(C(S'))$. But then the new language would be *more* efficient than the old, for we could copy an entire string from one regular location to another in just 1 step—by copying to an indexed location, and then back. Again, the issue here is what we would like our difficulty function to be.

These complications are caused by lookup charges. Then why not eliminate them entirely? That is, imagine a world in which looking something up is free, but charges are still made for printing and erasing. This could be achieved, for example, by retaining the present five classes of commands, but changing the difficulties for each of the last three classes to one. Consider, in this version, subroutine 15. Its difficulty will now be $L(C(C(S')))*L(C(C(C(S'))))$. Thus, a lookup charge has crept back in: It is reflected in the factor $L(C(C(S')))$, which arises from the necessity to store the string $C(C(S'))$ to implement this subroutine. It seems unnatural to have a lookup charge in this case but not in others. We could eliminate that charge here with a new basic command: SET $C(S) = C(C(C(S')))\ \{L(C(C(C(S'))))\}$. But then how will we deal with SET $C(S) = C(C(C(C(S'))))$? Again, a lookup charge will arise if this is made a subroutine, rather than an additional basic command. Are there examples in which such exotic indexed arrays actually impact the final difficulty functions?

Here is a more systematic method by which we might find a natural language with a natural difficulty function. We introduce *machine language(2)*, as follows: Storage locations are labeled by strings of exactly two characters, and each such location always contains exactly one character. Thus, $C(h8)$ denotes the character in location $h8$, whereas $C(C(h8)C(21))$ denotes the character in the location described by the two-character string whose first character is $C(h8)$ and whose second character is $C(21)$. In this machine language(2) there are (in addition to INPUT and OUTPUT, with which we are not concerned right now) four commands:

1. SET $C(xy) = z$
2. SET $C(C(xy)C(zw)) = C(pq)$
3. SET $C(pq) = C(C(xy)C(zw))$
4. IF $(C(xy) == z)$ SKIP N LINES

where x, y, z, w, p, and q are to be replaced by arbitrary explicit characters and n by an arbitrary (positive or negative) explicit integer. You can convince yourself that this is enough to carry out simple computations: manipulate strings (whose characters are now stored in individual locations), utilize indexed arrays, branch, count, etc. Indeed, machine language(2) is the *actual* machine language of my (very) old Apple II+. There are 256 characters, and thus the total RAM of the computer is just over 65 kB! The good news about machine language(2) is that there is an obvious choice of what difficulty to assign to each command: one step. The bad news is that machine language(2) cannot compute any problem at all (as we have defined those terms), for it utilizes a finite total memory. You can make available more memory by passing to machine language(3)—essentially the same as machine language(2), except that now *three* characters are needed to describe a location, with the obvious modifications of the basic commands listed earlier—or, if still more space is needed, to machine language(4), etc.

The idea, now, is the following: We would introduce a certain basic language, much like that of the previous section, together with a compiler, which would compile programs written in that language into machine language(n) for some n. (Indeed, this is what the Apple II+ does: Here, $n = 2$, and the basic language is Basic.) Given an input string S, the program that is actually run would be the compiled one, written in machine language(n). In this way, we obtain an unambiguous count of steps. If, in the course of that run, it emerged that more memory was needed, then the compiler would kick in again, to recompile the basic-language program in machine language(n') for some $n' > n$. Computation in machine language would then continue. It is best if these recompilations could take place seamlessly (e.g., if the machine-language commands could be adjusted so as to be n-universal). Thus, we are free to introduce any sorts of exotic commands we wish in our basic language—the only burden being that these be compiled into machine language. Moreover, we need not make hard choices as to what the difficulties of these commands are to be: They are whatever follows from their execution in machine language. Thus, since it is the machine language that assigns the difficulties, we might hope that those assignments will be the natural ones. Of course, it would still be required that we decide how to compare number of steps as carried out by machine language(n) with number as carried out by machine language(n'), for $n' \neq n$. It might be interesting to see whether this scheme could be implemented.

In any case, let us imagine that there has been introduced a natural language, which gives meaning to "algorithm," as well as an assignment of difficulty to each command in that language, which gives meaning to "difficulty of that algorithm."

Perhaps the major challenge in this subject is to obtain good lower limits on the difficulty-functions for computing various problems. That is, one would like to have theorems of the following form: "There is no program that computes this problem π and has difficulty function f satisfying $f \ll$ (something)." Here, "something" is some explicit difficulty function of interest.

Consider, as an example, multiplication of integers. The elementary multiplication that we all learned in school has difficulty function $\tilde{f}(S) = mn + 1$, where m and n are the numbers of digits of the two numbers. (This follows, since, in the course of the multiplication, each digit of the first number must be multiplied by each digit of the second.) Note that there do exist programs that multiply integers, with difficulty function $f \leq mn + 1$, and $f \not\sim mn + 1$. (Such a program, for example, might first check to see whether both integers are integral powers of 10, in which case it writes the product immediately; otherwise, it multiplies the numbers in the usual way.) But does there exist a program that computes this problem, with difficulty function $f \ll mn + 1$? It turns out that there does [7]. It is an open question, as far as I am aware, whether there exists a program (difficulty-function f) for multiplication of integers, such that there exists no program with difficulty function $f' \ll f$.

A more famous example is the prime problem, the problem that, given an integer n, returns the prime factors of that integer. The naive program that computes this problem (by trying each integer up to $n^{1/2}$ to see whether it divides n) has just over $n^{1/2}$. It is known that there are much more efficient programs. What, however, we do not have is a theorem that sets a good lower limit on the difficulty for *any possible* computation of this problem.

This key fact—that we lack good lower limits on the difficulty of computing various problems—has far-reaching implications for the structure of this subject. For instance, as we shall see later, there are examples of problems for which the use of quantum mechanics seems to allow a very efficient computation. In these examples, in particular, it appears that quantum mechanics is more efficient than any known regular computation of that problem. However, we cannot *prove* that quantum mechanics is more efficient, for we cannot eliminate the possibility that there exists some—miraculously efficient—regular computation that, for some reason, we have not yet discovered.

Chapter 13

Probabilistic Computing

As a prerequisite to our study of quantum-assisted computing, we consider here briefly the case of an ordinary computer that has access to a "random number generator." That is, we consider computing in a context in which the actions of a computer, at various stages during its operation, are subject to probabilities. Our purpose here is merely to understand how computing works in this environment. This will allow us, later, to separate effects arising from the full structure of quantum mechanics from those arising solely from its probabilistic character.

Consider the following programming language. The commands are precisely the six introduced in Sect. 11, except for the following change: The third command (APPEND) is replaced by

3. APPEND $x, y, \cdots, z$ TO $C(S)$ $\{1\}$

Here, $x, y, \cdots z$ stand for any finite list (possibly with repetitions) of explicit single characters from our character set and, as before, S stands for any explicit string. A program in this new language is defined just as before (i.e., as a finite list of commands, beginning with an INPUT and ending with an OUTPUT). Whenever, during the running of such a program, one of these new APPEND commands is reached, then whatever string is stored in location $C(S)$ is to be replaced by that same string, but with one of the characters $x, y, \cdots, z$ appended on the right. Which character is appended is to be selected randomly (i.e., with equal probability for each of the characters in the list). Thus, if there are $n \geq 1$ characters in the list, then each of those characters has probability of $1/n$ of being appended. Except for this one change, the programs run just as before. Note that our earlier programming language is a special case of this one, namely, that in which each APPEND command involves a list of exactly $n = 1$ character. We shall call this new language and its programs "probabilistic," when we wish to distinguish them from the original language and its programs. This new language is in the spirit of Turing machines and of the old language: We introduce the minimum that is necessary to get the job done.

Clearly, by allowing repetitions of characters in our list, we can achieve any rational-valued probability distribution for which character is appended to $C(S)$. Moreover, combining this new command with the others, we can achieve any rational probabilities for deleting (as opposed to not deleting) a character from the string in any given storage location, as well as rational probabilities for skipping various numbers of lines. Why *rational* probabilities? Why do we not simply allow arbitrary probabilities for appending various characters? This puts us on dangerous ground. Suppose, for example, that we allowed a command that appends x to a string, with probability c (the noncomputable number of Sect. 7) and appends y with probability $1 - c$. Armed with this command, we could write a probabilistic program to compute (in a sense we shall make precise in a moment) the halting problem! In short, we use only rational probabilities to prevent sneaking unauthorized information into the program through exotic choices of the probability numbers.

So, let us fix a probabilistic program and an input string. What can be the result if we run this program on this string? The possibilities in this case are precisely the same as before: The program can halt, with some output string, or it can continue forever without halting. But now, of course, different runs (with the same program and input string) can give different results. Denote by $\tilde{\mathcal{S}}$ the set consisting of all strings over our original character set $\mathcal{C}$, together with one additional element "$*$," which we designate "not halt." Then we can describe the running of a given program on a given input string by means of a probability distribution on $\tilde{\mathcal{S}}$. That is, for each $\alpha \in \tilde{\mathcal{S}}$, we have a nonnegative number $p(\alpha)$, called the "probability of outcome α," and these satisfy $\sum_{\alpha \in \tilde{\mathcal{S}}} p(\alpha) = 1$.

The following example will illustrate these ideas.

> **Example.** Consider the program begins by flipping a coin (i.e., applying an APPEND command with $n = 2$). If the coin comes up "heads," the program reports the total number of coin flips it has carried out (in this case, 1) and halts. If the coin is "tails," the program flips the coin again. Again if "heads" comes up, it reports the total number of flips (now 2); if "tails" it flips again. The program continues in this way. The possible outcomes in this example are the positive integers, together with $*$. The probability distribution is the following: For n a positive integer, $p(n) = 2^{-n}$; and $p(*) = 0$.

Note that, in this example, we have $p(*) = 0$ even though it is possible that a given run of this program will never halt. It turns out, however, that this phenomenon can occur only for this special outcome. We claim that, for any $\alpha \neq *$, $p(\alpha) > 0$ if and only if α is a possible outcome of running the program. The "only if" is immediate. For "if," let $\alpha \neq *$ be a possible outcome. This means that there exists a sequence of allowed steps in our program that ends with the program halting, with output string α. There must be only a finite total number of steps in this sequence (since the sequence ends up with a halt) and so a finite number of APPEND steps. Let r denote the (rational) number that results from multiplying the probabilities associated with the given passage through each of these APPEND steps. Then, clearly, $p(\alpha) \geq r > 0$.

> **Exercise.** Find an example of a probabilistic program such that the probability of some outcome (say, halting, with output the empty string) is the noncomputable number c of Sect. 7.

Now fix a probabilistic program, and also a problem, $\mathcal{S} \xrightarrow{\pi} \mathcal{S}$. We say that this program (*probabilistically*) *computes* problem π provided that, for every string S, the probability distribution resulting from running this program on initial string S satisfies the following: The probability of failing to halt is zero; and the probability of output string $\pi(S)$ is greater than the probability of every other output string. Think of this definition as requiring that we can extract $\pi(S)$ by "repeated running of the program on S." (We shall make this more precise shortly.) Note that $p(\pi(S))$ can be very small: We only require that no other individual string have probability greater than or equal to that of $\pi(S)$.

Clearly, every (nonprobabilistically) computable problem is also probabilistically computable, since every nonprobabilistic program is already a probabilistic program (namely, one in which each APPEND command happens to have but a single choice). It turns out that the converse is also true: Every probabilistically computable problem is also (nonprobabilistically) computable. In other words, the introduction of probability adds nothing to what can be computed.

To prove this, fix a problem π and a probabilistic program $\mathcal{P}$, that computes it, in the same sense as before. We now construct a nonprobabilistic program, $\tilde{\mathcal{P}}$, as follows: Given any input string S_0, this $\tilde{\mathcal{P}}$ simulates the action of $\mathcal{P}$ on S_0. That is, $\tilde{\mathcal{P}}$ keeps track, at each step, of which program line $\mathcal{P}$ is currently executing and what string resides in each of

$\mathcal{P}$'s storage locations. Then $\tilde{\mathcal{P}}$ simply follows the action of $\mathcal{P}$, step by step. When $\mathcal{P}$, so simulated, reaches an APPEND command, there will in general be several options for the next state (corresponding to the several possible characters that could be APPENDed in response to this command). When this happens, $\tilde{\mathcal{P}}$ simply keeps track of each of these options separately and also keeps a record of the probability for each. Thus, for example, if the simulation by $\tilde{\mathcal{P}}$ reaches the command APPEND x, y, z TO $C(k8)$, then $\tilde{\mathcal{P}}$ will consider separately the cases in which x, or y, or z is appended to the string $C(k8)$, assigning probability $1/3$ to each. Then $\tilde{\mathcal{P}}$ will simply simulate the action of $\mathcal{P}$ separately for each of the three cases. This branching continues for subsequent APPEND commands: If, say, one of these branches reaches another APPEND command, then there will result further branches (with new probability assignments) for $\tilde{\mathcal{P}}$ to follow.

Now, as $\tilde{\mathcal{P}}$ continues to follow all these branches, there will occur, every so often, a branch on which $\mathcal{P}$ would have encountered an OUTPUT command, and thus would have halted. When $\tilde{\mathcal{P}}$ reaches this point of a branch, then, of course, it can no longer follow that branch, since $\mathcal{P}$ itself would be unable to continue to operate along that branch. The program $\tilde{\mathcal{P}}$ maintains a table in which there is recorded, for each such terminated branch, two pieces of information: the $\mathcal{P}$ output string at that $\mathcal{P}$-halt, and the probability (a rational number) of reaching that particular termination. As $\tilde{\mathcal{P}}$ continues its simulation new terminating branches will be found, and this table will continue to grow. Note that the sum of the probabilities in this table will always be less than or equal to one. It further follows, from our assumption that $p(*) = 0$, that this sum will approach one, as $\tilde{\mathcal{P}}$ continues to run in this way.

Each time the program $\tilde{\mathcal{P}}$ adds a new line to this table, it will also perform the following calculation: First, it finds that string, S, having the largest total probability (i.e., that string such that the sum of the probabilities already assigned to that string in the table is greater than the sum of the probabilities already assigned to any other string). Then, $\tilde{\mathcal{P}}$ computes how much probability remains (i.e., it computes the number given by subtracting, from one, the sum of all the probabilities listed in the table). Next, $\tilde{\mathcal{P}}$ asks whether there is any other string, S', such that, if all the remaining probability were allocated to S' (in addition to the probability already assigned to S'), then this total would exceed the probability for S. If the answer to this question is yes (i.e., if there does exist a string S' having the potential of ultimately accumulating more probability than has already been assigned to S), then $\tilde{\mathcal{P}}$ continues to run. But eventually $\tilde{\mathcal{P}}$ must reach a point at which the answer to this question is no. That is, it will reach a point at which some string S has already accumulated enough probability that no other string is even a candidate ever to accumulate more. [This follows from the fact that the program $\mathcal{P}$ computes the problem π, i.e., that $p(*) = 0$ and $p(\pi(S_0))$ exceeds the probability of every other string.] When this happens, $\tilde{\mathcal{P}}$ itself halts and announces the winning string, S.

Clearly, this (nonprobabilistic) program $\tilde{\mathcal{P}}$ also computes the problem π. What we have shown, then, is that, given a probabilistic program that probabilistically computes a problem, we can, using simulation, build a nonprobabilistic program that computes the same problem[5]. In short, the use of probability adds nothing to the concept of computability.

We now turn to the issue of assigning a difficulty function to a probabilistic computation.

Fix a probabilistic program $\mathcal{P}$, that computes a problem π [so, for any input string S, the probability that $\mathcal{P}$ fails to

[5] In fact, this remains true even with certain, even weaker, notions of "probabilistically computable." For example, it suffices to require, instead of $p(*) = 0$, merely that we are given a program that, for each input string, computes the number $p(*)$. We also remark that one can modify this simulation program $\tilde{\mathcal{P}}$ to prove the following: If, for any probabilistic program acting on any string, $p(*)$ is computable, then each of the probabilities for each of the possible output strings is also a computable number.

halt for this string is zero, and the most likely output string is $\pi(S)$]. Fix any input string S, and let us run the program $\mathcal{P}$ on that string. Then during this run, various commands will be executed, and to each of these we have assigned a difficulty. Let us keep track of the cumulative total difficulty during the running of the program. Now should it happen, on this particular run of $\mathcal{P}$, that the program fails to halt, then the cumulative difficulty will, of course, grow without bound. But if $\mathcal{P}$ does halt, then there will be some total cumulated difficulty, ν, as of that halt. On different runs, there will be different cumulated difficulties. Thus we shall have some probability distribution on the possible cumulated difficulties; that is, for each ν, we have a number $p(\nu) \geq 0$, such that $\sum_\nu p(\nu) = 1$. [That this sum must actually be one follows from $p(*) = 0$.] Denote by $D(S)$ the mean total difficulty: $D(S) = \sum_\nu \nu\, p(\nu)$. This $D(S)$ is the difficulty that would be experienced "on the average" in one run of $\mathcal{P}$ with the given input string S. Of course, it is only an average: On any given run, it is entirely possible that the actual cumulated difficulty turn out to be much greater than $D(S)$—or much less. Note that the sum defining $D(S)$ need not converge: The difficulty ν could grow very quickly, even a $p(\nu)$ approaches zero.

> **Exercise.** Find an example in which the sum defining $D(S)$ diverges.

If this should occur, then we assign $\mathcal{P}$ infinite difficulty for the input string S, and we abandon further efforts to assign a difficulty function for this program. Note that, by simulating the running of $\mathcal{P}$ on input string S, as described above, we could compute an increasing sequence of rational numbers that converges to $D(S)$ [or, in the case in which $D(S) = \infty$, that grows without bound]. It seems unlikely, nevertheless, that D is always computable, in the sense that there always exists a (regular) program that, given probabilistic program $\mathcal{P}$, string S, and a positive integer n, returns a rational within $1/n$ of

$D(S)$. Indeed, even the problem of whether or not $D(S)$ is finite is probably not computable.

In any case, we have the notion of the mean difficulty, $D(S)$, for one run of $\mathcal{P}$ with input string S. But, unfortunately, a single run of this program, with input string S, does not tell us what the answer to our problem π is for the string [i.e., it does not tell us what $\pi(S)$ is]. Rather, we must run the program a number of times, on the same given input string, and keep a record of the various outputs. The "real" answer will be buried in the statistics of these records (in the form of the "most likely" output). What me must determine, then, is by what factor to multiply the mean difficulty, $D(S)$, to correct for this probabilistic character. To this end, let us run this program a total of r times, keeping a record of the various outputs that result. At the end of all these runs, we announce as the answer that output that occurred most frequently. Sometimes we will announce the correct answer, $\pi(S)$, and sometimes the wrong answer. Denote by $\kappa(r)$ the probability that our announcement is wrong. The following lemma states, roughly speaking, that, as the number r of runs increases, this probability $\kappa(r)$ goes to zero as e^{-Kr}, for a certain number K:

> **Lemma.** Consider a collection of positive numbers, with sum one. Denote the largest by p and the next largest by p', with $p > p'$. Carry out r runs in the corresponding probability distribution, and denote by $\kappa(r)$ the probability that the most frequent single outcome is not the most probable outcome (i.e., not the p-outcome). Then the limit of $[-\log \kappa(r)/r]$, as $r \to \infty$, exists, and has value $K = (p-p')^2/2[p(1-(p-p'))^2 + p'(1+(p-p'))^2]$.

The proof uses three facts: (i) For large r, any other outcome, say with probability $p'' < p'$, has negligible probability (compared with that of p') of being the most frequent outcome; (ii) the difference between the numbers of p-outcomes and p'-outcomes is, for large r, normally distributed, with mean $r(p - p')$ and squared variance $r[p(1 - (p - p'))^2 + p'(1 + (p - p'))^2]$; and (iii) the error function, $\text{erf}(x)$, satisfies $\lim_{x\to\infty}\text{erf}(x)/x^2 = -1/2$. For the present application, the p of the lemma is $p(\pi(S))$, and the p' is the probability of the next-most-likely outcome. Note that the number K of the lemma here depends on the input string S (through the dependence of the probabilities p, p' on S).

Now fix a small number $p_0 > 0$, which we shall interpret shortly as a confidence limit, that is, as the "largest probability of error that we are willing to tolerate, in our determination $\pi(S)$." Fix the input string S. Let us now run our program r_0 times, keeping track of the outputs for each run, and report as the answer that outcome that occurs most frequently in these runs. We wish to choose n_0 sufficiently large that the probability that this procedure results in the wrong answer does not exceed our confidence limit p_0. It follows from the lemma that, at least for sufficiently small p_0, the choice $r_0 \geq -(\log p_0)/K$ suffices, where K is the expression given in the lemma. That is, carrying out $r_0 > -(\log p_0)/K$ runs and reporting the most frequent outcome will, with probability at least $1 - p_0$, result in reporting $\pi(S)$. Note that, as we expect, the number of runs required grows without bound as $p_0 \to 0$.

We now take, as the difficulty of computing $\pi(S)$ using the probabilistic program $\mathcal{P}$ on input string S, the number $[-(\log p_0)/K]D(S)$, that is, the product of the minimum number (r_0) of runs required and the mean difficulty ($D(S)$) per run. Repeating this procedure for all possible input strings S, we obtain the difficulty function $f(S) = -(\log p_0)D(S)/K$ for this probabilistic computation of the problem π. But note that the confidence limit p_0 appears only in an overall factor. Thus, up to equivalence, it may be omitted. That is, it makes no difference how small a confidence limit p_0 we choose: The resulting difficulty function, up to equivalence, is independent of p_0. There results $f(S) = D(S)/K$. However, as it stands, this expression is not suitable for a difficulty function, because it is not in general bounded away from zero. [Indeed, as $p \to 1$ (whence $p' \to 0$), $K \to \infty$.] The reason for this phenomenon is quite simple: Our argument requires, in the limit $p \to 1$, that the program $\mathcal{P}$ be run only a small fraction of one time! But at least one full run of $\mathcal{P}$ is necessary in any case, and we can take this fact into account by adding $D(S)$ to the $f(S)$ above. Doing this, and passing to an equivalent difficult function, we obtain the following:

> Let $\mathcal{P}$ be a probabilistic program that computes some problem, π. Then we shall assign to this program the difficulty function given by $f(S) = D(S)(p + p')/(p - p')^2$, where $D(S)$ is the mean difficulty for running $\mathcal{P}$ on input string S, p is the probability that that run results in output $\pi(S)$, and $p' < p$ is the probability of the next most probable output.

The factor, $(p+p')/(p-p')^2$, by which $D(S)$ is multiplied reflects the increase in difficulty because $\mathcal{P}$ computes our problem only probabilistically. This factor is always at least one, and for $p = 1$ (and so $p' = 0$) this factor is exactly one (i.e., the difficulty function for a probabilistic program reduces, in the special case of a nonprobabilistic program, to our original difficulty function). When p and p' are very close, the factor is large, reflecting the fact that many runs of $\mathcal{P}$, on the given input string S, must be carried out to have a reasonable chance of announcing the correct value of $\pi(S)$.

We have already seen that the introduction of probability adds nothing to what is computable. But can probability add to efficiency? Consider the following assertion:

> **Assertion.** Let $\mathcal{P}$ be a probabilistic program that probabilistically computes problem π, with difficulty function f. Then there exists a nonprobabilistic program $\mathcal{P}'$ that also computes π, and whose difficulty function f' satisfies $f' \le f$.

This assertion states, in other words, that any probabilistic program can always be at least matched, in terms of efficiency, by a corresponding nonprobabilistic program. It seems likely, intuitively, that this assertion is true. Given string S, there is some definite string $\pi(S)$ that you wish to compute. Why would it ever be more efficient to use a random number generator to find this $\pi(S)$?

One might imagine that one could prove this assertion by using a simulation, as described earlier. Given probabilistic program $\mathcal{P}$ that computes π, then we can, by simulating $\mathcal{P}$, construct a nonprobabilistic program $\tilde{\mathcal{P}}$ that also computes π. If we could show that the difficulty of $\tilde{\mathcal{P}}$ is always less than or equal to that of $\mathcal{P}$, then we would be done. Unfortunately, this is false in general.

> **Example.** Let π be the problem that assigns to each string (represented as a positive integer) the string "a". Consider the probabilistic program $\mathcal{P}$ that operates as follows: Given input n, $\mathcal{P}$ flips a coin n times, and then simply rolls a die. If the die comes up 1, $\mathcal{P}$ reports "b"; otherwise, $\mathcal{P}$ reports "a". This $\mathcal{P}$ computes the problem π and has difficulty function f given by $f(n) = n$. Denote by $\tilde{\mathcal{P}}$ the nonprobabilistic program that simulates $\mathcal{P}$, as described earlier. Thus, $\tilde{\mathcal{P}}$ also computes π. But, because of the 2^n branches created by $\mathcal{P}$ (via its coin flips), this $\tilde{\mathcal{P}}$ has difficulty function $\tilde{f}(n) = 2^n$. Thus, $f \ll \tilde{f}$.

Thus, the program $\mathcal{P}$ merely creates 2^n branches and then proceeds to ignore them! Clearly, there is no need, in this example, for $\tilde{\mathcal{P}}$ to follow all 2^n branches: They are all the same, so it suffices for $\tilde{\mathcal{P}}$ to follow a single branch. It is obvious, in this example, how $\tilde{\mathcal{P}}$ can avoid unnecessary computations. But to prove the assertion we need to find a general way for $\tilde{\mathcal{P}}$ to do this. For example, the probabilistic program $\mathcal{P}$ could actually ignore many branches but could be so written to disguise this fact from $\tilde{\mathcal{P}}$. Thus, if such a simulation is ever going to work to produce a proof of the assertion, then it will be necessary to design a "smart" simulation program $\tilde{\mathcal{P}}$—one that looks ahead to find the more promising branches. That this idea can be implemented is by no means obvious. Thus, it is not at all clear whether or not there exists a problem, along with a probabilistic program that computes that problem, that stands as a counterexample to the assertion-above.

Amazingly enough, this assertion remains open! Even an example of a problem, together with a probabilistic computation of that problem, such that it appears plausible that there is no nonprobabilistic computation that is at least as efficient, would be most interesting. Here is an possible strategy to obtain such an example.

Fix a problem π_0 that accepts as input any pair of positive integers, (N, k), with $k \leq N$, and produces as output either "yes" or "no". Thus, for each value of N a total of N values of k are allowed $(1, \cdots, N)$. Further, let this π_0 have the following property:[6] For each N, either (i) $\pi_0(N, k) =$ "yes" for at least two-thirds of the allowed k values or (ii) $\pi_0(N, k) =$ "no" for at least two-thirds of the allowed k-values. From this π_0, we construct a new problem, π, acting on positive integers N, as follows: $\pi(N)$ is "yes" or "no", according as whether case (i) or (ii) applies for that N. This π is the problem of interest.

Now let there be given a program $\mathcal{P}_0$ that computes π_0. Then we can easily write a program $\mathcal{P}$ that computes π: Given any positive integer N, let the program $\mathcal{P}$ simply run $\mathcal{P}_0$ on (N, k) for each of the allowed k-values, $k = 1, \cdots, N$, and report "yes" or "no" according to which answer resulted more frequently in those N runs. (Actually, it is enough to run $\mathcal{P}_0$ for just over two-thirds of these k-values.)

[6] This property is imposed in a very strong form, for ease of exposition. It can be weakened considerably.

Denote by f_0 the difficulty function of $\mathcal{P}_0$. For each positive integer N, denote by $g(N)$ the maximum value achieved by $f_0(N,k)$, as k runs through $1,\cdots,N$. Let us also suppose that, for each fixed N, all of the $f_0(N,k)$, for $k=1,\cdots,N$, are between, say, $g(N)/2$ and $g(N)$. (This supposition is made merely to simplify the discussion; it could be weakened considerably.) Then, up to equivalence, the difficulty function of the program $\mathcal{P}$ (since it merely runs $\mathcal{P}_0$ a total of N times) is $Ng(N)$.

Here is a probabilistic program $\mathcal{P}_{\text{prob}}$, that also computes the problem π. For each positive integer N, let $\mathcal{P}_{\text{prob}}$ select, randomly, a positive integer $k \leq N$, run the program $\mathcal{P}_0$ on the pair (N,k), and report whatever $\mathcal{P}_0$ reports on this single run. Note that this probabilistic program $\mathcal{P}_{\text{prob}}$ always halts, and that, for each N, the probability of its giving the correct answer is at least two-thirds. Thus, this $\mathcal{P}_{\text{prob}}$ does indeed probabilistically compute π. The probabilistic program $\mathcal{P}_{\text{prob}}$ has difficulty function given simply by $g(N)$. That is, the probabilistic program $\mathcal{P}_{\text{prob}}$ is *far* more efficient than the nonprobabilistic program $\mathcal{P}$.

So, it would appear to be relatively easy to find an example of a problem for which the probabilistic program is more efficient than the nonprobabilistic one. All we must do is find a problem π_0 of the type just described. Here is a simple example: Let $\pi_0(N,k)$ be "yes" if k is a prime and "no" if k is composite. Then this π_0 (with suitable adjustments for the first few N values) satisfies the "two-thirds" condition above. In this example, then, for given N, the nonprobabilistic program $\mathcal{P}$ must check for primeness of each integer $k=1,\cdots,N$, whereas probabilistic program $\mathcal{P}_{\text{prob}}$ checks for primeness of single (randomly selected) integers in this range. Clearly, $\mathcal{P}_{\text{prob}}$ is far more efficient than $\mathcal{P}$. But in this case there exists a shortcut for computing the problem π—there exists a nonprobabilistic program that is much more efficient than either of these two. This is the program that (except possibly for the first few N-values) always reports "no". Thus, the probabilistic program $\mathcal{P}_{\text{prob}}$ in this example, although more efficient than $\mathcal{P}$, is not more efficient than *every* nonprobabilistic program that computes π. This example, in other words, fails.

What is needed, then, is a computable problem π_0 that reports, for each N, either "yes" for at least two-thirds of the allowed k-values $(1, \cdots, N)$ or "no" for at least two-thirds of the allowed k-values, but which is such that there is no shortcut for determining which of these answers is the more frequent. It must be the case that the only way to determine whether two-thirds of the answers are "yes" or two-thirds "no" is actually to compute $\pi_0(N, k)$ for the requisite number of k-values (or, at least, to do some equally difficult computation). Remarkably enough, no such example of a π_0 seems to be known! A theorem to the effect that there exists no such π_0 (i.e., a theorem to the effect that this type of example will always fail) would be extremely interesting.

For most probabilistic programs of interest, the probability of the correct outcome dominates the other probabilities, and when this is the case the formula we used above can be simplified. Suppose that there exists a number $a > 0$, independent of S, such that $p[\pi(S)]$ exceeds the next-highest probability by at least this a. This condition is always satisfied, for example, if all the $p[\pi(S)]$ are greater than 0.501. Indeed, one could take the position that $\mathcal{P}$ is not "really computing" the problem unless this condition is satisfied. In any case, whenever this condition is satisfied, then the factor $(p+p')/(p-p')^2$ in our formula is bounded above, and so in this case the difficulty function, up to equivalence, is given simply by $D(S)$. That is, under this rather weak condition, we may assign to a program $\mathcal{P}$ that solves the problem π the difficulty function whose value on any input string is the mean difficulty of running that program on that input string.

Exercise. Let $\mathcal{P}$ and $\mathcal{P}'$, with respective difficulty functions f and f', both compute the same problem. Is there a way to alternate between these two programs, constructing a program $\mathcal{P}''$ that also computes this problem, with difficulty function given by $f''(S) = \min[f(S), f'(S)]$?

The probabilistic programs that one would typically write will have the property that, for every run of the program on every input string, the program will always halt. [As we have seen from an earlier example, this property is stronger than merely requiring that $p(*) = 0$.] An example would be a Monte Carlo program: It visits an APPEND command with several outcomes a certain number of times; keeps track of the results, thus generating a distribution of outcomes; and then simply halts, reporting, say, some property of this distribution. For such a program—one that always halts, on every run with every input string—the previous discussion can be simplified considerably. Fix such a program $\mathcal{P}$ and an input string S. We claim first that, in this special case, there is but a finite number of possible outcomes. To see this, call a state, during the running of this program, *rich* if there is an infinite number of possible outcomes starting from that state. Suppose, for contradiction, that the initial state—the original INPUT of the initial string—were rich. Now follow the steps of this program as it runs. As long as we encounter only non APPEND commands, we must always remain in a rich configuration. Now consider the first APPEND command we encounter. We are already in a rich state at this point, and so at least one of the possible branches from this point must itself be rich. Choose any such rich branch. Continuing in this way (taking, at each probability APPEND command, some choice that again results in a rich configuration) we will always remain in a rich configuration, and therefore we can never halt (since the halt state is hardly rich). But this contradicts our assumption that failure to halt, for any run on any string, is not possible. This shows that our initial supposition—that there was an infinite number of possible outcomes—must be false. A similar argument shows that, in this situation (a program, running on an input string, such that the failing to halt is not a possible outcome), there must be an upper bound to the number of steps that will ever be required to achieve the halt. The proof is the identical to that just given; we merely redefine "rich" as "having no upper bound for the maximum number of steps that could be required, from that point, to achieve the halt." These two proofs are essential the same as that of the "tree theorem" in mathematics. It also follows, in this case, that the probabilities for the different output strings are all rational (since there is a finite number of possible routes to a given output string, and each of these, since it encounters a finite number of APPEND commands, has a rational probability of being the actual route). And finally, the simulation program $\tilde{\mathcal{P}}$ that we introduced earlier will, in this case, halt all by itself (without invoking the special rule involving a probability calculation). This follows, since each of the branches in the simulation must, eventually, be terminated.

Thus, many of the complications of simulating the running of a probabilistic program, and of computing its difficulty function, disappear in the special case that the program, for every run with every input string, always halts.

Chapter 14

Quantum Mechanics

This section is a very short course in quantum mechanics—for people who already know quantum mechanics.

A *Hilbert space* is a complex vector space, equipped with an inner product that is antilinear in the first factor and linear in the second, such that the associated norm is positive-definite. All our Hilbert spaces will be finite-dimensional.[7] Vectors in Hilbert spaces are usually written, for example, as $|\alpha\rangle$, where α is some symbol or word that describes the vector. and the inner product of vectors $|\alpha\rangle$ and $|\beta\rangle$ is usually written $\langle\alpha|\beta\rangle$. *The* states *of a quantum system are described by nonzero vectors (up to an overall complex factor) in a suitable Hilbert space.*

Let H and H' be Hilbert spaces. The *tensor product* of H and H' is a certain Hilbert space obtained by taking linear combinations of formal products, where each product is of one vector in H with one vector in H'. The tensor product is written $H \otimes H'$ and has dimension given by the product of the dimensions of H and H'. For $|\alpha\rangle \in H$ and $|\alpha'\rangle \in H'$, the corresponding formal product, in $H \otimes H'$, is written

[7]The full definition of a Hilbert space includes an additional condition of completeness, but in the finite dimensional, and in that case completeness follows automatically.

$|\alpha\rangle|\alpha'\rangle$. *Now consider two quantum systems, whose states are described by respective Hilbert spaces* H *and* H'. *Regard these two separate systems as one. Then the Hilbert space of states of the combined system is* $H \otimes H'$. *Indeed,* $|\alpha\rangle|\alpha'\rangle$ *represents that state of the combined system with the* H *system in state* $|\alpha\rangle$ *and the* H' *system in state* $|\alpha'\rangle$. Since the Hilbert space $H \otimes H'$ allows linear combinations of these simple products, not every state of the combined system is one in which each of the original systems is in a particular state.

An *operator* on a (finite-dimensional) Hilbert space H is a linear mapping from H to itself. For example, the identity, I, is an operator, as is, for any $|\alpha\rangle \in H$, the map, written $|\alpha\rangle\langle\alpha|$, with action $|\alpha\rangle\langle\alpha|\ (|\beta\rangle) = |\alpha\rangle\ (\langle\alpha|\beta\rangle)$. For A an operator and $|\alpha\rangle$ a vector in the Hilbert space, we sometimes write $|A\alpha\rangle$ for $A(|\alpha\rangle)$. For A and A' operators on Hilbert spaces H and H', respectively, we write $A \otimes A'$ for the operator on $H \otimes H'$ with action $(A \otimes A')(|\alpha\rangle|\alpha'\rangle) = |A\alpha\rangle|A'\alpha'\rangle$ (extended to all of $H \otimes H'$ by linearity). We shall sometimes not distinguish between an operator A' acting on H' and the operator $I \otimes A'$ acting on $H \otimes H'$.

An operator U on a Hilbert space is called *unitary* if it is inner-product preserving (i.e., if $\langle U\alpha|U\beta\rangle = \langle\alpha|\beta\rangle$ for every α, β). For example, if $|\alpha\rangle$ is unit, then $I - 2|\alpha\rangle\langle\alpha|$ is unitary. *The evolution of a quantum system through time is described by a unitary operator* U*: Initial state* $|\psi\rangle$ *evolves to* $|U\psi\rangle$.

An operator on a Hilbert space is called *Hermitian* if it satisfies $\langle A\alpha|\beta\rangle = \langle\alpha|A\beta\rangle$ for every α, β. For example, I and $|\gamma\rangle\langle\gamma|$ are Hermitian. In the finite-dimensional case, every Hermitian operator has a finite number of eigenvalues, all real, and the corresponding eigenspaces span the entire Hilbert space. *Observations on quantum systems are described by Hermitian operators. Let a system, initially in the state given by unit* $|\psi\rangle$, *be observed via Hermitian* A. *Then the "result" of the observation is one of the eigenvalues of* A, *the state of the system after the observation is the projection of* $|\psi\rangle$ *into the corresponding eigenspace, and the probability of that result is the squared norm of that projection.* Given a basis for H, by an observation *via that basis* we mean an observation via a Hermitian operator whose eigenspaces are those generated by the individual basis vectors.

Chapter 15

Grover Construction

We now begin a new subject: quantum-assisted computing. Our strategy will be first to consider, in some detail, one particular example. We shall then generalize. We choose for our example what is called the Grover construction [4, 10, 11], for it has a number of attractive features: It is very simple, it illustrates most of the constructs and ideas of quantum-assisted computing, and it holds out realistic hope of generating an example in which the quantum assist provides a genuine reduction in difficulty.

Consider the challenge of finding a needle in a haystack. Fix an integer N (which you should think of as containing, say, 100 digits). The haystack is the set consisting of the N integers $0, 1, \cdots, (N-1)$, and the needle is a specific one of those integers, say k_0. We suppose that we have a computer that allows us to search for the needle in the following manner: The computer accepts as input any integer k with $0 \leq k \leq (N-1)$, and returns either "no" (if $k \neq k_0$) or "yes" (if $k = k_0$). We wish to find the needle. The obvious way to do this is to run the computer for various k-values as input. Thus, to be certain of finding k_0 we would have to run the computer a total of N times, whereas a mere 50% chance would require only $N/2$ runs. The issue is whether we can discover a way to find the needle in substantially fewer runs.

Here is a corresponding quantum system. Let there be given an N-dimensional Hilbert space, H_{in}, with orthonormal basis $|0\rangle, |1\rangle, \cdots |N-1\rangle$. This is the quantum system in which the input will be registered. In addition, let there be given a two-dimensional Hilbert space, H_{out}, with orthonormal basis $|\text{no}\rangle, |\text{yes}\rangle$, to register the output. Then the Hilbert space with which the computer (and we) interact is $H_{\text{in}} \otimes H_{\text{out}}$. We represent the action of the computer by the following unitary operator[8] on this Hilbert space:

$$V(|k\rangle|\text{no}\rangle) = |k\rangle|\text{no}\rangle \; (k \neq k_0), \qquad V(|k_0\rangle|\text{no}\rangle) = |k_0\rangle|\text{yes}\rangle, \quad (4)$$

$$V(|k\rangle|\text{yes}\rangle) = |k\rangle|\text{yes}\rangle \; (k \neq k_0), \qquad V(|k_0\rangle|\text{yes}\rangle) = |k_0\rangle|\text{no}\rangle. \quad (5)$$

That is, if the input register is in any state other than $|k_0\rangle$, then V does nothing, whereas if it is in state $|k_0\rangle$, then V flips the output state. This unitary operator V is a reasonable rendition of what a computer might do. Indeed, suppose we have agreed to start the system with the output register in state $|\text{no}\rangle$. Then Eq. (4) specifies that V records the correct answer (for the given $|k\rangle$) in H_{out}, and Eq. (5) is the simplest way to extend this V, as a unitary operator, to all of $H_{\text{in}} \otimes H_{\text{out}}$.

Let us pause at this point to see how we might search for the needle under this setup. First select any candidate k, then begin with the registers in the corresponding initial state, $|k\rangle|\text{no}\rangle$, and then run the computer (i.e., apply V). When the computer is finished [with final register-state that given in Eqs. (4) and (5)], make an observation, on H_{out}, via the basis $|\text{no}\rangle, |\text{yes}\rangle$. If the result is "no" [which it will be, with probability $(N-1)/N$], then we know that our trial k was not the needle; while if it is "yes" (probability $1/N$) then we have found our k_0. This will be recognized as merely the original search, cloaked in a thin veneer of quantum mechanics.

Let us now change things slightly. Set $|\phi\rangle = \frac{1}{\sqrt{N}}(|0\rangle + |1\rangle + \cdots + |N-1\rangle)$, a unit vector in H_{in}. This is a state that

[8]The action of V on the linear combinations of these simple product states is, of course, fixed by linearity.

combines all possible inputs, equally weighted. Let us now begin with state $|\phi\rangle|\text{no}\rangle$. Then the running of the computer produces

$$V(|\phi\rangle|\text{no}\rangle) = \frac{1}{\sqrt{N}}\{|0\rangle + \cdots + |k_0 - 1\rangle + |k_0 + 1\rangle + \cdots + |N-1\rangle\}\ |\text{no}\rangle + \frac{1}{\sqrt{N}}|k_0\rangle|\text{yes}\rangle.$$

Again, let us see what we can learn from this final state. We first make an observation on H_{out} via its basis. With probability $(N-1)/N$ we will obtain "no," in which case we have learned nothing whatsoever (not even, as in the previous paragraph, a k known not to be the needle). But, one time out of N, we will get lucky and obtain "yes." In this case, we proceed to make an observation on H_{in} via its basis $|0\rangle, |1\rangle, \cdots, |N-1\rangle$. The result (since now the H_{in}-state is simply $|k_0\rangle$) will tell us what k_0 is. But note that even this procedure, using the state $|\phi\rangle \in H_{\text{in}}$, has not gained us anything: This is still basically the original search, the only essential difference being that now quantum mechanics is "choosing" our trial k's for us.

Let us now make still another change, this time to the output register. Let us now choose as our initial state $|\phi\rangle\ \frac{1}{\sqrt{2}}\{|\text{no}\rangle - |\text{yes}\rangle\}$. In this case, the running of the computer produces

$$V\left(|\phi\rangle\ \frac{1}{\sqrt{2}}\{|\text{no}\rangle - |\text{yes}\rangle\}\right) = \frac{1}{\sqrt{N}}\ \{|0\rangle + \cdots + |k_0 - 1\rangle - |k_0\rangle + |k_0 + 1\rangle + \cdots + |N-1\rangle\}\frac{1}{\sqrt{2}}\ \{|\text{no}\rangle - |\text{yes}\rangle\}.$$

That is, the output register is now always in the state $\frac{1}{\sqrt{2}}\{|\text{no}\rangle - |\text{yes}\rangle\}$—both before and after the running of the computer. All the computer does, now, is reverse of the sign of the $|k_0\rangle$-term in the input register. What can we learn by making our observations on *this* final state? Absolutely nothing. An observation on H_{out}, via its basis, will give equal probability for "no" and "yes," and an observation on H_{in}, via its basis, will return each $k = 0, 1, \cdots, (N-1)$ with equal probability. It looks as though we have gone backward.

Undaunted, we set $V_{\text{in}} = I - 2|k_0\rangle\langle k_0|$, a unitary operator (reflection across the plane orthogonal to $|k_0\rangle$) on H_{in}. Then the result of the previous paragraph can be summarized as follows: For any $|\psi\rangle \in H_{\text{in}}$,

$$V\left(|\psi\rangle\ \frac{1}{\sqrt{2}}\{|\text{no}\rangle - |\text{yes}\rangle\}\right) = |V_{\text{in}}\psi\rangle\ \frac{1}{\sqrt{2}}\{|\text{no}\rangle - |\text{yes}\rangle\}.$$

That is, provided the H_{out}-state is set to $\frac{1}{\sqrt{2}}\{|\text{no}\rangle - |\text{yes}\rangle\}$, the action of V (the run-the-computer operator) on $H_{\text{in}} \otimes H_{\text{out}}$ is represented by the action of this V_{in} on H_{in}, with the H_{out}-state never changing. Next, set $W = I - 2|\phi\rangle\langle\phi|$, another unitary operator (reflection across the plane orthogonal to $|\phi\rangle$) on H_{in}. Note that W does *not* involve knowing which $|k\rangle$ is the needle in the haystack. We now have, by an easy calculation,

$$-WV_{\text{in}}|\phi\rangle = \frac{N-4}{N}\ |\phi\rangle + \frac{2}{\sqrt{N}}\ |k_0\rangle. \tag{6}$$

Thus, we are now working solely in H_{in}, for we begin with H_{out}-state $\frac{1}{\sqrt{2}}\{|\text{no}\rangle - |\text{yes}\rangle\}$, and this state never changes. Equation (6) gives the result of starting with state $|\phi\rangle \in H_{\text{in}}$, then running the computer (i.e., applying unitary V_{in}), and then applying unitary W.

Again, let us pause to interpret this equation. Let us make an observation, on the state given by the right side of (6), via our basis, $|0\rangle, |1\rangle, \cdots, |N-1\rangle$, for H_{in}. We find (taking the inner product of that right side with $|k_0\rangle$ and squaring the result) that the probability of obtaining k_0 is $(3N-4)^2/N^3$ (the rest of the probability being distributed equally over the other k's). For large N, this probability is about $9/N$. After observing via this basis (obtaining a k-value), we may of course check directly, by running our classical computer, whether that k is actually the needle. Nine times out of N, we will in this way find the needle. Note that this is nine times the *a priori* probability of finding k_0 by merely guessing a k-value. It may look as though we are making some real progress here, but this appearance is misleading. Even a factor of 9 in the probability for success still means that, in order to find the needle, we must carry out a number of runs proportional to N. But suppose that, instead of observing the state (6) immediately, we repeat the operation. Apply $-WV_{\text{in}}$ again, and only then observe via the $|k\rangle$ basis and check the k-value that results? Our probability of success will then turn out to be 25 times the *a priori* probability. These remarks motivate what follows.

Now comes the key step: to look, from a geometrical viewpoint, at what we have just done. Consider the 2-plane in H_{in} spanned by $|k_0\rangle$ and $|\phi\rangle$. Each of the operators of interest, V_{in} and W, when acting on any vector orthogonal to this 2-plane, is the identity. Thus, all the action is taking place within this 2-plane. Let us choose an orthonormal basis for this 2-plane consisting of $|k_0\rangle$ and $|k_0\rangle^\perp$, where the latter is that linear combination of $|k_0\rangle$ and $|\phi\rangle$ that is unit and orthogonal to $|k_0\rangle$. Denote by θ the angle that $|\phi\rangle$ makes with $|k_0\rangle^\perp$. Then $\sin\theta = \langle k_0|\phi\rangle = \frac{1}{\sqrt{N}}$.

Now, each of V_{in} and $-W$ is a certain reflection within this plane (about vectors $|k_0\rangle^\perp$ and $|\phi\rangle$, respectively). But the composition of two reflections in a plane is a rotation. The angle of rotation is given by cos(angle) $= \langle\psi|(-WV_{\text{in}})\psi\rangle$, where $|\psi\rangle$ is any unit vector in our 2-plane. Choosing $|\psi\rangle = |\phi\rangle$ (or $|k_0\rangle$, if you prefer), we find that this angle is precisely 2θ.

So, vector $|\phi\rangle$ starts out making angle θ with $|k_0\rangle^\perp$, and each application of $-WV_{\text{in}}$ increases that angle by 2θ. So, if we apply $-WV_{\text{in}}$ to $|\phi\rangle$ a total of s times, the resulting vector will make angle $(2s+1)\theta$ with $|k_0\rangle^\perp$. Now apply the operator $-WV_{\text{in}}$ to $|\phi\rangle$ a total of s times, where s is that number, such that $(2s+1)\theta$ is closest to $\pi/2$. Then this number of times will satisfy $s \le \pi/(4\theta) \le (\pi/4)\sqrt{N}$, where in the second inequality we used $\theta \ge \sin\theta = \frac{1}{\sqrt{N}}$. Having applied

$-WV_{\text{in}}$ to $|\phi\rangle$ this many times, there results vector in this plane within angle θ of $|k_0\rangle$. Let us now make an observation on this final vector, via the $|k\rangle$-basis for H_{in}. The probability that this observation results in k_0, by what we just observed, is greater than or equal to $\cos^2\theta = 1-\frac{1}{N}$. That is, our chances are excellent that this single observation on H_{in} will find the needle.

So, to summarize, if we apply, to initial state $|\phi\rangle\ \frac{1}{\sqrt{2}}\{|\text{no}\rangle - |\text{yes}\rangle\}$ in $H_{\text{in}} \otimes H_{\text{out}}$, the operator $-WV$ a number of times not exceeding $\frac{\pi}{4}\sqrt{N}$, and then observe the resulting state via the $|k\rangle$-basis, we will, with probability at least $1 - \frac{1}{N}$(i.e., almost certainly), obtain the needle, k_0. Note that we only have to run the computer (i.e., apply V) a number of times proportional to $\sqrt{N}$—not to N itself. It does indeed appear that there has been a significant gain in efficiency. This is an example of a quantum-assisted computation.

Note that if you are impatient—insisting on making $|k\rangle$-basis observations between the computer runs (just to see how things are going), then you will destroy this effect. This is similar to the familiar "watched pot never boils" parable in quantum mechanics.

Chapter 16

Grover Construction: Six Issues

In the previous section, we gave an example of a construction that appears to show quantum mechanics providing a clear gain in efficiency over a nonquantum computation. We discuss here six issues pertaining to that construction.

16.1 Initial State

The construction requires that the registers be placed, initially, in state $|\phi\rangle \frac{1}{\sqrt{2}}\{|\text{no}\rangle - |\text{yes}\rangle\}$. Is it feasible to build this state?

The state of H_{out} would not seem to be much of a problem: After all, this is merely a two-dimensional Hilbert space. So, for example, we could represent this space physically as the spin states of a spin-1/2 particle, designating $|\text{no}\rangle$ and $|\text{yes}\rangle$ as the states corresponding to the spin aligned or antialigned in a given direction. Then $\frac{1}{\sqrt{2}}\ \{|\text{no}\rangle - |\text{yes}\rangle\}$ would be the state in which the spin is aligned in a certain orthogonal direction.

However, the state $|\phi\rangle \in H_{\text{in}}$ is more complicated. After all, this is a superposition of N states. To construct these states one at a time, and then "superpose them" (whatever that means), is a job that threatens to have difficulty N

(i.e., to overwhelm the difficulty of running the computer $\sqrt{N}$ times). Here is a device—common in this subject—to circumvent this problem. Fix, once and for all, a two-dimensional Hilbert space, with basis $|0\rangle, |1\rangle$ (not to be confused with the vectors of the same name in H_{in}). So, for example, this H might be the spin states of a spin-1/2 system. Let $N = 2^n$ for some positive integer n. (To achieve this—at most a doubling of the number of input states—should not cause too much additional complication.) Now set

$$H_{\text{in}} = H \otimes H \otimes \cdots \otimes H, \tag{7}$$

where a total of n copies of H appear on the right.[9] Note that this gives the correct dimension for H_{in}. Now consider a typical state, for example, $|0\rangle|1\rangle|1\rangle|0\rangle \cdots |0\rangle|1\rangle$ (total of n factors) in the Hilbert space on the right. We identify this with the state $|k\rangle$ of H_{in}, where $k = 0110 \cdots 01$ in base 2. The k's that result in this way range from 0 (for $00 \cdots 0$) to $2^n - 1$ (for $11 \cdots 1$), so we indeed obtain in this way the basis we want for H_{in}. Under this identification, the construction of the state $|\phi\rangle \in H_{\text{in}}$ is quite easy: It is a simple product

$$|\phi\rangle = \frac{1}{\sqrt{2}}(|0\rangle + |1\rangle)\ \frac{1}{\sqrt{2}}(|0\rangle + |1\rangle) \cdots \frac{1}{\sqrt{2}}(|0\rangle + |1\rangle)$$

of the states $\frac{1}{\sqrt{2}}\ (|0\rangle + |1\rangle)$ for each of the H-factors. This follows by expanding the right side and using the definitions of $|k\rangle$ and $|\phi\rangle$. Thus, with this choice of how H_{in} is to be structured, the construction of the state $|\phi\rangle$ should be relatively easy. We note that this construction could have been carried out[10] with any fixed dimension for the factor-Hilbert spaces H.

---

[9]Note that we could not, for example, let the H's be simply the spin states for n identical spin-1/2 particles, because the states on the right in (7) are not in general antisymmetric under particle interchange. However, we could, for example, have a system of n electrons occupying n energy levels (say, in an atom), where each H, referring some energy level, gives the spin state of the occupant of that level.

[10]In fact, the different H's in the product could, if we so desired, be assigned different dimensions. Exercise. Set up a system of arithmetic

16.2 Final Observation on H_{in}

The construction requires that, at the end of the computer runs, an observation be made on H_{in} via the $|k\rangle$-basis. Is it feasible to make such an observation?

Yes, it is. Denote by A the Hermitian operator $|0\rangle 0\langle 0| + |1\rangle 1\langle 1|$ on H (so observation of A is observation of H via its natural basis). For example, if H is spin-states, and the basis is spin-component in a certain direction, then A would be the observation of the spin component in that direction, plus 1/2. Now consider the following Hermitian operator on $H \otimes \cdots \otimes H$: Operator $(2^{n-1}A)$ applied to the first H-factor, plus operator $(2^{n-2}A)$ applied to the second H-factor, and so on, until reaching finally operator $(2^0 A)$ applied to the last H-factor. (Here, we are regarding these operators on the H-factors as operators on the tensor product in the manner described in Sect. 14.) The resulting sum can, via Eq. (7), be regarded as an operator on H_{in}, and we note that it does indeed have the $|k\rangle$ as its eigenstates. (In physical terms, observe the first H-component and multiply by 2^{n-1}, observe the second and multiply by 2^{n-2}, etc., and add. The result will be precisely the k-value of that state.) We would expect to have no difficulty in making an observation of H via this operator A and, therefore, no difficulty in making an observation of H_{in}, so constructed, via its $|k\rangle$-basis.

16.3 Building the Operator W

The construction requires that we apply unitary operator $W = I - 2|\phi\rangle\langle\phi|$ to H_{in}. Is it feasible to build and apply such an operator?

Note that this by no means follows immediately from the prior point: The mere fact that we feel capable of placing H_{in}

---

in which each of the various digits of an integer refers to a different number-system base. Figure out how to add in this system (which turns out to be quite simple!).

in state $|\phi\rangle$ does not lead directly to an interaction on H_{in} that shifts each state $|\psi\rangle \in H_{\text{in}}$ to state $W|\psi\rangle \in H_{\text{in}}$. To build the operator W, we proceed as follows.

We first require a few preliminaries. We introduce a more convenient basis for H: $|\alpha\rangle = \frac{1}{\sqrt{2}}(|1\rangle+|0\rangle)$, $|\beta\rangle = \frac{1}{\sqrt{2}}(|1\rangle-|0\rangle)$. In terms of this basis, we have $W = I - 2|\alpha\rangle\cdots|\alpha\rangle\langle\alpha|\cdots\langle\alpha|$. Next, we introduce a unitary operator T on $H \otimes H \otimes H$, with the following action: $T(|\alpha\rangle|\alpha\rangle|\alpha\rangle) = |\alpha\rangle|\alpha\rangle|\beta\rangle$, $T(|\alpha\rangle|\alpha\rangle|\beta\rangle) = |\alpha\rangle|\alpha\rangle|\alpha\rangle$, with T the identity on the other six basis elements of $H \otimes H \otimes H$. That is, this operator T, which is called the *Toffoli gate*, flips the third H-state if and only if the first two H-states are[11] both $|\alpha\rangle$; so, for example, we have $T^2 = I$. Let us denote by $H_1, H_2, \cdots, H_n$ the n H's in the tensor product that is H_{in}. We now introduce a second Hilbert space, $H_{\text{scratch}} = \tilde{H}_3 \otimes \tilde{H}_4 \otimes \cdots \otimes \tilde{H}_{n+1}$, where each of the $\tilde{H}$'s in this tensor product is also a copy of our basic Hilbert space H. This is the Hilbert space in which we shall carry out scratch work. Thus, our full Hilbert space is now $H_{\text{in}} \otimes H_{\text{scratch}}$, a tensor product of $2n-1$ copies of H. Now consider the following operator on this tensor product:

$$\begin{aligned}\mathcal{W} = {} & T(H_n, \tilde{H}_n, \tilde{H}_{n+1})T(H_{n-1}, \tilde{H}_{n-1}, \tilde{H}_n)\cdots T(H_3, \tilde{H}_3, \tilde{H}_4) \\ & \times T(H_1, H_2, \tilde{H}_3). \end{aligned} \tag{8}$$

We note that this operator, as a composition of unitary operators, is unitary. Let us now begin with an arbitrary state in H_{in}, but with H_{scratch} in the state $|\tau\rangle = |\beta\rangle|\beta\rangle\cdots|\beta\rangle \frac{1}{\sqrt{2}}(|\alpha\rangle - |\beta\rangle)$. Let us apply to this state the operator (8), and see what happens. The rightmost operator T in this composition will place $\tilde{H}_3$ (which began in state $|\beta\rangle$) in state $|\alpha\rangle$ if and only

[11] The general state in $H \otimes H \otimes H$ is, of course, not one in which each of the H's is in a particular state ($|\alpha\rangle$ or $|\beta\rangle$); rather, it is a superposition of all eight possible combinations of individual H-states. We often describe operators, such as this T, by giving their actions on each of the combinations that appear in this superposition. Thus, when we say, for example, "the first two H-states are ...," we really mean "that term, in the superposition, in which the first two H-states are ..."

if the H_1- and H_2-states are both $|\alpha\rangle$. The next T, reading from right to left, will place $\tilde{H}_4$ in state $|\alpha\rangle$ if and only if H_3 and $\tilde{H}_3$ are both in state $|\alpha\rangle$ (i.e., if and only if H_1, H_2, and H_3 are all in state $|\alpha\rangle$). Similarly, the next T will place $\tilde{H}_5$ in state $|\alpha\rangle$ if and only if all four of H_1, H_2, H_3 and H_4 are in state $|\alpha\rangle$. Continue in this way, working from right to left in (8). Recall that the last $\tilde{H}$, $\tilde{H}_{n+1}$ begins in state $\frac{1}{\sqrt{2}}(|\alpha\rangle - |\beta\rangle)$ rather than $|\alpha\rangle$. Thus, in the last step, an attempt to "flip" the H_{n+1}-state will merely introduce a minus sign. We conclude: The operator $\mathcal{W}$ of (8), acting on a state $|\psi\rangle|\tau\rangle \in H_{\text{in}} \otimes H_{\text{scratch}}$, where $|\psi\rangle$ is any state in H_{in}, and $|\tau\rangle$ is the state in H_{scratch} given above, indeed generates a sign change if all the H's are in state $|\alpha\rangle$, and no sign change otherwise.

The operator $\mathcal{W}$, so constructed, is our candidate for W. Of course, it acts, not merely on the Hilbert space H_{in} (as W does), but rather on $H_{\text{in}} \otimes H_{\text{scratch}}$. Nevertheless, it does seem to have the right action and so, it appears, would seem to suffice for the Grover construction.

However, this appearance is misleading. The candidate $\mathcal{W}$ will not work as a proxy for W, for the following reason: Some scratch work for this calculation was left in the auxiliary Hilbert space H_{scratch}. That is, the final state, after application of $\mathcal{W}$, is an entanglement of H_{in}-states and H_{scratch}-states. Consider, for example, $n = 4$. If the initial state of H_{in} was $|\alpha\rangle|\alpha\rangle|\beta\rangle|\alpha\rangle$, say, then the final state of H_{scratch} will be $|\alpha\rangle|\beta\rangle\ \frac{1}{\sqrt{2}}(|\alpha\rangle - |\beta\rangle)$, whereas if the initial state of H_{in} was $|\alpha\rangle|\beta\rangle|\beta\rangle|\alpha\rangle$, then the final state of H_{scratch} will be $|\beta\rangle|\beta\rangle\ \frac{1}{\sqrt{2}}(|\alpha\rangle - |\beta\rangle)$. This entanglement, we claim, will destroy the working of the Grover construction. To see this, consider Eq. (6), which gives the result of the first application of $-WV_{\text{in}}$ to $|\phi\rangle$: a rotation $|\phi\rangle$ through angle 2θ. The key to the construction is that the next application of $-WV_{\text{in}}$ (as well as each successive application) must rotate through an additional angle 2θ. But, for this to happen, there must occur cancellation between the $|\phi\rangle$'s and $|k_0\rangle$'s that arise from application of $-WV_{\text{in}}$ to the two terms on the right in (6).

Now consider what happens if the W on the left in (6) is replaced by $\mathcal{W}$. Then the terms on the right side of this equation will become entangled with various elements of H_{scratch}. Therefore, on the next application of $-WV_{\text{in}}$ the necessary cancellations on the right will not take place. The Grover construction will thus fail.

In order to obtain an effective W, we proceed as follows: Set

$$\mathcal{W}' = T(H_3, \tilde{H}_3, \tilde{H}_4) \cdots T(H_{n-1}, \tilde{H}_{n-1}, \tilde{H}_n)\, \mathcal{W}. \tag{9}$$

That is, $\mathcal{W}'$ first applies $\mathcal{W}$, and then applies all the operators of $\mathcal{W}$, save the leftmost, in reverse order. It is easy to check that this procedure undoes the entanglement. That is, we have $\mathcal{W}'(|\psi\rangle|\tau\rangle) = |W\psi\rangle|\tau\rangle$ for any $|\psi\rangle \in H_{\text{in}}$, where $|\tau\rangle \in H_{\text{scratch}}$ is the initial state given earlier. *This* $\mathcal{W}'$, then, can be used in place of W in the Grover construction.

We conclude, then, that the operator W on H_{in} in the Grover construction can indeed be built, by introducing an auxiliary Hilbert space H_{scratch} and applying the Toffoli gate (an operator on $H \otimes H \otimes H$) a total of $2n - 3$ times. So, it would seem that the operator W is feasible—provided the operator T is feasible. We shall return to this last issue shortly.

16.4 Building the Operator V

No real computer, it might be argued, operates by applying some unitary operator V to $H_{\text{in}} \otimes H_{\text{out}}$, as in the Grover construction. After all, real computers use irreversible operations (such as placing bits in locations). How, then, are we to construct and interpret the operator V?

Here is a model for how a computer might operate. We introduce an additional Hilbert space, H_{com}, to represent the computer states. Then the total Hilbert space is $H_{\text{in}} \otimes H_{\text{out}} \otimes H_{\text{comp}}$. The running of the computer will then be represented by some unitary operator $\mathcal{V}$ on this Hilbert space. Let us fix a vector $|\psi_{\text{init}}\rangle \in H_{\text{comp}}$ to represent the initial state of the computer. Then, in the Grover case (i.e., with H_{in} spanned by $|0\rangle \cdots |N-1\rangle$ and H_{out} by $|\text{no}\rangle, |\text{yes}\rangle$), the action of a suitable $\mathcal{V}$ would be as follows:

$$\mathcal{V}(|k\rangle|\text{no/yes}\rangle|\psi_{\text{init}}\rangle) = |k\rangle|\text{no/yes}\rangle|\psi_k\rangle, \tag{10}$$

where the H_{out}-state on the right is $|\text{no}\rangle$ or $|\text{yes}\rangle$ depending on whether the H_{out}-state on the left is $|\text{no}\rangle$ or $|\text{yes}\rangle$, and also on whether or not $k = k_0$. The $|\psi_k\rangle \in H_{\text{comp}}$ on the right in Eq. (10) is the final state in which the computer finds itself, depending on the k-value on the left (and also on the choice of initial state in H_{out}, which we suppress). This operator $\mathcal{V}$ is unitary, and so it is invertible. Thus, we are suggesting, the operation of any computer must always be reversible. [Indeed, in a world governed by quantum mechanics, this is necessary, for dynamics therein is described by an (invertible) unitary operator.] Things do not appear to be this way in practice only because we fail to take into account how large and complicated H_{comp} can be. It includes not only the states of the chips, wires, fan, etc. within the box but also (if, say, the computer is plugged in) the states of the electric company, and then of its employees, etc. By the time all this dust settles, things look pretty irreversible.

Unfortunately, the operator $\mathcal{V}$ of (10) will not serve as a proxy for the operator V of the Grover construction. The problem is that $\mathcal{V}$ introduces entanglements between $H_{\text{in}} \otimes H_{\text{out}}$ on the one hand and H_{comp} on the other, as reflected in the dependence of the final computer state, $|\psi_k\rangle$, in (10) on k. These entanglements, in the same manner as for $\mathcal{W}$ in the previous discussion, will interfere with the cancellation that must take place in Eq. (6), and will thereby cause the Grover construction to fail. To avoid these entanglements, we must, for example, so design our computer that the final computer state, say $|\psi_{\text{final}}\rangle$, is independent of $|k\rangle$. Then, when it comes time to repeat the computation, we could either apply some special treatment to the computer to restore its initial state to $|\psi_{\text{init}}\rangle$, or discard that computer entirely, bringing in another with the initial state $|\psi_{\rm init}\rangle$ already preinstalled.[12] It would be best if we could arrange that $\mathcal{V}$ automatically, at the end of each run, returns the computer to state $|\psi_{\rm init}\rangle$, ready for the next run.

So, in any case, to carry out the computation implicit in the Grover construction, we shall have to produce a computer that does not introduce entanglements between computer states and in–out states. This is definitely *not* the computer on your desk! We shall have to build our computer anew. The danger we face is that the building and operating of such computers consumes resources—in particular, time—and we must be careful that this consumption does not overwhelm the apparent savings we derive from using quantum mechanics.

Recall that the Hilbert space $H_{\rm in} = H \otimes \cdots \otimes H$, in the Grover construction, has large dimension, 2^n. Our computer must interact with this large Hilbert space but do so relatively efficiently. It would be of great help if we could design our computer to interact, not with all n of the H's at once, but rather with only a few at a time. Does this restriction entail a restriction on the possible unitary operators we can generate on $H_{\rm in} \otimes H_{\rm out}$? The following shows that it does not.

> **Theorem.** Let H be a finite-dimensional Hilbert space. Then any unitary operator on $H \otimes H \otimes \cdots \otimes H$ is equal to a product of unitary operators, each of which acts on at most two of the H-factors in this tensor product.

Of course, different combinations of the two H-factors are allowed for the different unitary operators in this product.

[12] Why not get rid of these awkward entanglements, not by searching for a clever $\mathcal{V}$, but rather by simply discarding the computer after each run, bringing in a new computer, with $|\psi_{\rm init}\rangle$ preinstalled, for the next run? The problem with this maneuver is that the act of discarding a system entangled with another places the latter system in a mixed state, as described by a density operator. But a mixed state for $H_{\rm in}$ destroys the cancellation, and so the Grover construction, as surely as does entanglement.

Our proof of the theorem will make use of three facts.

Lemma 1. Every Hermitian operator on $H \otimes \cdots \otimes H$ is a linear combination of operators of the form $A \otimes \cdots \otimes B$, where $A, \cdots, B$ are Hermitian operators on H.

Lemma 2. Every Hermitian operator on a Hilbert space is a linear combination of commutators of Hermitian operators, and the identity I.

Lemma 3. Fix a connected Lie group G and a collection of one-parameter subgroups of G. If the generators of these subgroups generate the entire Lie algebra of G, then the subgroups themselves generate the entirety of G.

Lemma 1 is easy to prove by a dimensional argument, using the fact that the dimension of the (real) vector space of Hermitian operators on a Hilbert space is equal to the square of the dimension of that (complex) space. Let there be n H's in the tensor product, each of dimension m. Then the Hilbert space $H \otimes \cdots \otimes H$ has dimension m^n, and so the vector space of Hermitian operators on this space has dimension $(n^m)^2$. The Hermitian operators of the form given in the lemma form a subspace of this space, and it has dimension (dimension of Hermitian operators on $H)^m = (n^2)^m$. These dimensions are equal, and so the subspace is the entire vector space.[13] Lemma 2 (which, apparently, is of little independent interest) follows by direct construction. For $m = 2$, for example, it is the statement that any linear combination of spin operators is a commutator of two such linear combinations. In Lemma 3, the Lie algebra of a Lie group is the tangent space at its identity element. The "generator" of a one-parameter subgroup is that element of the Lie algebra given by the tangent

[13] Lemma 1 is not quite as empty as it may appear at first sight. To see this, you might try to write the Hermitian operator that switches the two H states in $H \otimes H$ in the form guaranteed by the lemma.

to that curve at the identity. The Lie algebra "generated" by these generators is the collection of all elements that can be obtained by using linear combinations and brackets on the generators of the one-parameter subgroups. And, finally, that the elements of these subgroups "generate" the entirety of G means that every element of G can be written as a (finite) product of such subgroup elements. This lemma, in other words, states that if you can get the entire group from the subgroups "infinitesimally close to the identity," then you can indeed get the entire group from the subgroups "everywhere."[14] (This is the sort of thing that would normally be used, without mention, in a physics course.)

The theorem is very easy to prove from the three lemmas. Consider, say, $n = 3$. We have, for A, B, C, and D any Hermitian operators on H, and a any real number,

$$[A \otimes I \otimes C,\ I \otimes B \otimes D] + a\, A \otimes B \otimes I = A \otimes B \otimes ([C, D] + aI), \tag{11}$$

where [,] denotes i times the commutator. Each of the operators that appears on the left contains an I and so is a generator of unitary operators on $H \otimes H \otimes H$ that act on only two factors. By Lemma 2, the right side of (11) includes the general tensor product of three Hermitian operators on $H \otimes H \otimes H$; and, by Lemma 1, these span all Hermitian operators on the tensor product. The result (for $n = 3$) now follows from Lemma 3. The case of general n is by induction on n, repeating the construction of (11) at each step.

It seems likely that the product (whose existence is guaranteed by the theorem) involves no more than 5^{n-2} factors (perhaps substantially fewer), by an argument that traces the

[14] As an example, let G be the rotation group, let one one-parameter subgroup be the rotations about some vector $\vec{s}$, and let another be rotations about some other, independent, vector $\vec{t}$. Then, since the Lie bracket of the corresponding infinitesimal rotations is simply an infinitesimal rotation about $\vec{s} \times \vec{t}$, the hypothesis of Lemma 3 is satisfied. The lemma asserts in this case that *every* rotation can be written as a some product of various rotations about $\vec{s}$ and $\vec{t}$. This fact is the basis of Euler angles.

mechanism of Lemma 3. Unfortunately, this number grows quickly with n. The theorem is also true (suitably modifying Lemma 1) when the H's in the tensor product have different dimensions.

So, we may expect to build our computer out of operators that act on H-factors two at a time. But is it feasible to construct even these operators? Let us take, as an example, the case in which H is two dimensional, representing the spin states of an electron. Then the general Hermitian operator on H is $\vec{s} \cdot \vec{\sigma} + bI$, where $\vec{s}$ is any vector in 3-space, $\vec{\sigma}$ is the vector (Pauli) spin operator, and b is any real number. (Note that these form a four-dimensional vector space, as required.) The Hermitian operator $\vec{s} \cdot \vec{\sigma}$ generates the family of unitary operators, written $e^{i\,\vec{s}\cdot\vec{\sigma}}$, that correspond to rotations in space about the vector $\vec{s}$ as axis; bI generates the family, written e^{ib}, that corresponds to overall phase changes (which have no physical significance).

The unitary operators on a single H can be constructed physically as follows: The unitary operators corresponding to rotations about $\vec{s}$ result from applying to the electron a magnetic field in the $\vec{s}$ direction, for such a field causes the electron, by virtue of its angular momentum and magnetic moment, to precess about the magnetic-field direction. The product of the field strength and the time for which the interaction is turned on determine the magnitude of this rotation.

However, to invoke the theorem we must also construct the unitary operators on $H \otimes H$ (i.e., on the two-electron system). It should be clear that merely subjecting the two electrons, each to its own magnetic field, will not suffice. We must introduce some sort of direct interaction between the two electrons. One such is what is called the spin-spin interaction. The corresponding Hermitian operator on $H \otimes H$ is $\vec{\sigma_1} \cdot \vec{\sigma_2}$, where $\vec{\sigma_1}$ and $\vec{\sigma_2}$ denote the spin operator acting on the first and second factor in $H \otimes H$, respectively. (Strictly speaking, we should include a $\otimes$ between the two σ's in this expression; but the dot gets in the way.) This particular interaction actually occurs in nature: If the two electrons are merely brought close together,[15] then, by virtue of the electromagnetic interaction between their magnetic moments, the electrons interact in just the manner we have described. The corresponding unitary operator may be written $e^{ia\,\vec{\sigma_1}\cdot\vec{\sigma_2}}$, where the number a is determined by how close together the electrons are placed, and for how long.

That these two physical operations—placing one or both electrons in a magnetic field and allowing the electrons to interact electromagnetically—suffice to generate all possible two-electron interactions now follows from the following theorem:

> **Theorem.** Let H be a two-dimensional Hilbert space. Then every unitary operator on $H \otimes H$ is equal to some (finite) product of the operators e^{ib}, $I \otimes e^{i\,\vec{s}\cdot\vec{\sigma_1}}$, $I \otimes e^{i\,\vec{s}\cdot\vec{\sigma_2}}$, and $e^{ia\,\vec{\sigma_1}\cdot\vec{\sigma_2}}$, where $\vec{s}$ is any vector in 3-space and a and b are any real numbers.

The proof is virtually identically to that of the earlier theorem, where the lemmas are used in the same way. In this case, Eq. (11) is replaced by

$$-[\vec{t}\cdot\vec{\sigma_1}\otimes I,\ [\vec{s}\cdot\vec{\sigma_1}\otimes I,\ \vec{\sigma_1}\cdot\vec{\sigma_2}]]+(\vec{s}\cdot\vec{t})\,\vec{\sigma_1}\cdot\vec{\sigma_2} = (\vec{s}\cdot\vec{\sigma_1})\otimes(\vec{t}\cdot\vec{\sigma_2}), \tag{12}$$

where we have used the fact that $[\vec{s}\cdot\vec{\sigma},\ \vec{t}\cdot\vec{\sigma}] = i(\vec{s}\times\vec{t})\cdot\vec{\sigma}$. Taking linear combinations involving the right side of (12) and the Hermitian operators $\vec{s}\cdot\vec{\sigma_1}\otimes I$, $I\otimes\vec{s}\cdot\vec{\sigma_2}$, and $I\otimes I$ reproduces the entire Lie algebra of $H\otimes H$, which is just what we need to complete the proof.

Here are a couple of examples. Consider the operator W on $H_{\text{in}} = H\otimes\cdots\otimes H$, where H is taken as the two-dimensional Hilbert space of spin-1/2 states. It follows, from these two

[15]This could be done, for example, by keeping the electrons in boxes, with H representing the spin state of the occupant of a given box, and then moving the boxes into close proximity.

theorems, that this W is equal to a composition of our basic unitary operators: that (on a single H) generated by a magnetic field and that (on two H's) generated by the spin–spin interaction. It seems likely that the number of such basic operators that must be composed to construct W in this manner increases exponentially in n. Note that this construction of W is different from that of Sect. 16.3, for there we made use of an auxiliary Hilbert space H_{scratch}, whereas here there is none. Next, note, again from the two theorems, that the Toffoli operator T on $H \otimes H \otimes H$ is also equal to a product of the basic operators on H. Having constructed T from the basic operators, we may then proceed to construct W from the basic operators, using the strategy of Sect. 16.3. Although this alternative construction of W requires an auxiliary Hilbert space H_{scratch}, it does have the advantage that the number of basic operators required grows only linearly with n. We may, in addition to W, also construct V_{in}, in the following manner: Denote by U the unitary operator on H with action $U|0\rangle = |1\rangle, U|1\rangle = |0\rangle$. Suppose that the needle, written in base 2, is, say, $k_0 = 01001 \cdots 01$. Then set $\mathcal{U} = U \otimes I \otimes U \otimes U \otimes I \cdots \otimes U \otimes I$, where the U's and I's on the right correspond to the digits in this expression for k_0. We then claim that $V_{\text{in}} = \mathcal{U}W\mathcal{U}$. This is easy to check: $\mathcal{U}$ sends $|k_0\rangle$ to $|1\rangle|1\rangle \cdots |1\rangle$ (but other $|k\rangle$'s to something else), and then W produces a minus sign (but, for other $|k\rangle$'s, a plus sign); and then the final $\mathcal{U}$ restores the original state. Note that, in this construction of V_{in}, the number of operators that must be composed grows only linearly with n.

This discussion shows that there will in general be a variety of ways to construct a given unitary operator out of some set of basic operators. Some ways may involve an auxiliary Hilbert space (in which case we must avoid entanglement) and some not; some may involve the composition of a large number of basic operators and some a smaller number. But, unfortunately, none of this is what we really want for the Grover construction. The V_{in} here, for example, is completely useless for our purposes, because it requires that you already "know" k_0, and this is exactly what you are not supposed to know. What we need is, not a variety of ways to construct some "given" unitary operator on a Hilbert space out of basic operators, but rather a way to convert programs into operators. In the Grover case, for example, we begin with a computer program that checks whether or not a given k is the needle, and we wish to convert that program to a suitable unitary operator V_{in}. Here is a general summary of what we are looking for.

> Let there be given a program $\mathcal{P}$ that accepts as input nonnegative integers and returns nonnegative integers. We may register the inputs and outputs in Hilbert spaces H_{in} and H_{out}, each of which is a (finite) tensor product of some finite-dimensional Hilbert space H with itself. We wish to convert the program $\mathcal{P}$ into a unitary operator on $H_{\text{in}} \otimes H_{\text{out}}$, such that this operator "computes" the output from the input in the manner of $\mathcal{P}$ and does so with substantially the same difficulty function as that of $\mathcal{P}$. This unitary operator may require an auxiliary Hilbert space H_{scratch}, but if it does then the operator must be such that the $H_{\text{in}} \otimes H_{\text{out}}$-part of the final state is not entangled with the H_{scratch}-part.

It is not clear how to make this summary into a precise statement. What, for example, does "...in the manner of ..." mean, and what is the "difficulty function" of a unitary operator? We have in mind some sort of compiler, which turns command lines in the program into compositions of some basic unitary operators on the Hilbert spaces. But it is not clear how this is to work. What, for example, are the operator equivalents of APPEND and DELETE? Even more difficult would be finding an operator-equivalent of IF (LAST $C(S) == x$) THEN SKIP n LINES. In any case, having constructed such a compiler, then we might be able to define a suitable difficulty function in terms of the number of basic operators in the composition. And even after all this, we would have to contend with the fact that *programs* accept as input arbitrary integers, whereas our *operators* act on just finite tensor products. This situation is somewhat reminiscent of the issue, discussed in Sect. 13, of compiling programs in machine language. It might be worthwhile to try to resolve that issue first, as a prerequisite to this one.

There seems to be a sense, in this field, that there may exist some sort of construction along the lines just outlined.

16.5 Errors

Errors abound in the Grover construction. They can appear in the setting up of the initial state, in the application of the operators W and V, and in the observation on the final state. Errors could arise, for example, from imperfections in the apparatus, or from quantum tunneling causing interactions between the H-states and the thermal fluctuations in the outside world. How shall we take such errors into account?

Of course, errors abound everywhere in physics. But here, because of the kinds of questions we are asking, this issue seems particular compelling. Consider a computation, and suppose that, on the initial run, the input string S is such that we require just 10 steps. Then we can afford, for this run, to be relatively cavalier about errors. However, the next run, for another S, might involve 1,000 steps, requiring us to purchase new and better equipment just to keep the effects of errors in check. And still another run might involve 1,000,000,000 steps, requiring that we cool the entire Earth down to 0.03 K and move the Sun over to another part of the Galaxy. All of these extra precautions take time and effort, and so should be taken into account in the difficulty function. But how will we be able ever to include such things? How, for example, does the difficulty of these various precautions scale with the number of steps in the computation? Similar issues arise already in ordinary computing: Bits are sometimes recorded incorrectly, and the longer the calculation the greater the care that must be exercised in this regard.

We shall simply ignore the effects of errors, not out of any conviction these effects are likely to be unimportant, but rather because we do not know how to do anything else.

16.6 What Is the Problem?

The Grover construction, as you have undoubtedly noticed, does not compute any problem at all, at least not as we have defined that term. A "problem" entails an output for *every* possible input string, whereas the Grover construction searches for the needle in a finite haystack. Finite input sets are not very interesting: All problems based on them are computable, and all difficulty functions on them are equivalent.

The obvious way to respond to this situation would be to modify the construction to apply to a variety of haystacks. Given N, we first construct the Hilbert space H_{in} (as a tensor product of about $\log_2 N$ copies of a two-dimensional H), then build our computer (i.e, construct our operator V). We are now prepared to apply the Grover construction to find the needle. Of course, the building of the computer (i.e., of the operator V) is an additional burden, which would, presumably, be included in the difficulty function for this computation.

So let us suppose, then, that we have suitably modified the Grover construction along these lines. We would then be in a position to ask the key question: Is there any problem for which the Grover construction, so configured, is more efficient than any computation method not using quantum mechanics? Here is a precise mathematical assertion that reflects these ideas.

Assertion. Let π be any problem that accepts as input a pair (N, k), where N is a positive integer and k is one of the integers $(0, 1, \cdots, N-1)$, and returns either "yes" or "no", such that for each N, there is one and only one of those k's (call it k_0) for which $\pi(N,k)$ is "yes". Let π' be the problem that accepts as input any positive integer N and returns this k_0. Let $\mathcal{P}$ be any program (written, say, in the language of Sect. 12) that computes π, with difficulty function f; and set $h(N) = \max_k f(N,k)$. Then there exists a program $\mathcal{P}'$ that computes π', with difficulty function f' satisfying $f' \leq \sqrt{N}h$ (in the sense of difficulty functions).

The idea of this assertion is the following: Think of the problem π as a sequence of needle-in-the-haystack challenges. Here, N represents the haystack itself, and k is a needle candidate for that haystack. Thus, the problem π takes such a haystack and needle candidate and returns the answer to the following question: "Is that k the needle for that haystack?" The program $\mathcal{P}$ computes this π. That is, $\mathcal{P}$ tests whether, for a given haystack, a given candidate for the needle is indeed the correct needle (k_0). By contrast, the problem π' simply announces, given the haystack, what the needle is for that haystack. And the program $\mathcal{P}'$ computes that problem π'. That is, $\mathcal{P}'$ takes the haystack and finds the needle.

Note that, given a program $\mathcal{P}$ that computes π, we can immediately write a program $\mathcal{P}'$ that computes π': For each given N, this $\mathcal{P}'$ merely runs $\mathcal{P}$ on the pair (N,k), for each $k = (0, 1, \cdots, N-1)$ in turn. It then finds that k for which $\mathcal{P}$ returns "yes", and announces that k-value. (Thus, in particular, if π is computable, then automatically π' is.) What will be the difficulty function, f', of this program $\mathcal{P}'$? Fix N. Then the maximum difficulty to check a single k-value is the $h(N)$ given in the assertion. Since $\mathcal{P}'$ must run $\mathcal{P}$ at most N times, we have $f'(N) \leq Nh(N)$. We conclude that if, in the inequality of our assertion, $\sqrt{N}$ were replaced by N, then the assertion would be true, choosing for $\mathcal{P}'$ this program that simply runs $\mathcal{P}$ a total of N times.

But the assertion as it stands says that there always exists a shortcut $\mathcal{P}'$ that is *better*, in a suitable sense, than the naive $\mathcal{P}'$ we have just constructed. It says that you can always discover some way of finding the needle with difficulty not exceeding the maximum $\mathcal{P}$-difficulty to check one candidate, multiplied by the *square root* of the number of candidates that $\mathcal{P}'$ would have to check. The assertion asserts, in other words, that given any family of needle challenges, and a way to meet those challenges by trial and error, then there exists a way to meet those challenges that is more efficient than trial and error, by a factor of $\sqrt{N}$.

Why do we get this $\sqrt{N}$? It comes from the Grover construction! Imagine that we had somehow come up with a counterexample to our assertion. That is, we have a problem π, of the type just indicated, and program $\mathcal{P}$ that computes it, such that there does not exist any shortcut program P' in the sense of the assertion. Fix N, and consider the action of $\mathcal{P}$ on (N, k), for that N. Let us next imagine that we were able to simulate this action of $\mathcal{P}$ by a suitable unitary operator, V, on $H \otimes \cdots \otimes H$ (n times, where $2^n \geq N$) and that the total "difficulty" required to apply this operator was just $h(N)$ [i.e., the maximal difficulty that the ordinary program $\mathcal{P}$ encounters for the various (N, k), with this fixed N]. And finally, let us suppose further that the additional difficulty of building the quantum system is sufficiently small that it may be disregarded. Then the Grover construction would find the needle (with very large probability) with a total difficulty of $\sqrt{N}h(N)$ (since, as we saw in Sect. 15, the computer would have to be run only $\sqrt{N}$ times). But we began with the assumption that this $\mathcal{P}$ is a counterexample to the assertion [i.e., that it is such that there exists no shortcut $\mathcal{P}'$ with $f'(N) \leq \sqrt{N}h(N)$]. What this means, in other words, is that there is no regular program that solves this problem more efficiently than does the Grover construction.

We conclude that a counterexample to our assertion would provide a road map for finding (via Grover) an example in which a quantum-assisted computation is more efficient than any computation of the same problem without a quantum assist. If the assertion were true, however, then this result would considerably diminish the prospects for using the Grover construction in this way.

We emphasize that the assertion above is a statement in mathematics: It does not involve quantum mechanics nor any details of how computations work. It is either true or false. I have neither a proof nor a counterexample to this assertion. The following three examples of (failed) attempts to construct a counterexample are intended to give a sense of how the assertion works.

For the first example, let $\pi(N, k)$ be "yes" if and only if $k = N/2$ [or $(N+1)/2$, if N is odd]. Thus, program $\mathcal{P}$ would, on receiving (N, k), multiply k by 2, and see if the result is N (or $N+1$). The difficulty function is $\log(N)$ (i.e., effectively, the number of digits of N, independent of k); and so we have $h(N) = \log(N)$. This program $\mathcal{P}$ is not a counterexample. Let $\mathcal{P}'$ be the program that accepts positive integer N, simply computes $N/2$ or $(N+1)/2$, and returns that integer. The difficulty function for this $\mathcal{P}'$ is also $f'(N) = \log(N)$, and so we certainly have $f'(N) \leq \sqrt{N}h(N)$. This candidate for a counterexample was hopeless right from the beginning. Any time the structure of $\mathcal{P}$ is "do some computation involving N, and then check to see if the result matches k," then you will never end up with a counterexample. Program $\mathcal{P}'$ will overhear this strategy and proceed to compute the needle k_0 directly from N in the same manner, ending up with difficulty function given by $f'(N) = h(N)$, and thus satisfying the inequality of the assertion.

For the second example, let $\pi(N, k)$ be "yes" if and only if k is the largest prime $\leq (N-1)$. Thus, program $\mathcal{P}$ would, on receiving (N, k), first check to see whether k is prime (reporting "no" if it is not), then check the integers from $k+1$ to $N-1$ for primeness reporting "no" if any are prime, and, if none are, reporting "yes". The difficulty function $f(N, k)$ of this program has complicated k dependence. But in any case, denote by $h(N)$ the greatest difficulty encountered as k ranges from 0 to $(N-1)$. This program $\mathcal{P}$ is not a counterexample.

Let $\mathcal{P}'$ be the program that works downward from $N - 1$, checking each integer for primeness and reporting the first prime it finds. This $\mathcal{P}'$, then, computes the problem π'. The number of steps required by $\mathcal{P}'$ will be the same as for $\mathcal{P}$ to check candidate k_0 [i.e., we have $f'(N) = f(N, k_0)$]. Hence, $f' \leq h$, and so certainly the inequality of the assertion will be satisfied. This candidate for a counterexample was not much more promising. Any time the structure of $\mathcal{P}$ is "check to see if k is the largest integer $\leq (N - 1)$ such that ...," then $\mathcal{P}'$ will overhear this strategy and proceed to find the needle directly by starting at $(N - 1)$ and working downward. Similarly for "smallest", and any other "-est" that $\mathcal{P}'$ can figure out how to exploit.

The third example is the following: For N any positive integer, denote by p_N the integer obtained by writing out the digits of π (314159265...) and stopping as soon as you arrive at the largest integer less than N^2 (e.g., $p_{32} = 314$). Now let $\pi(N, k)$ be "yes" if and only if either (i) p_N is not the product of exactly two primes and $k = 0$ or (ii) p_N is the product of exactly two primes and k is the smaller prime factor of p_N. In other words, the needle, for haystack N, is the smaller prime factor of p_N if p_N is a product of exactly two prime factors, and it is 0 otherwise. [The idea here is that the digits of π are "pretty random" and that it is (or at least, used to be) hard to find prime factors other than by trial and error.] Now, most of the time (i.e., for most N) p_N will have many factors, and in these cases it will be easy to find the needle. Program $\mathcal{P}'$ will have a field day in these cases, easily achieving $f'(N) \leq \sqrt{N}h(N)$. But every so often (at least, we *hope* so we, of course, have no theorem to this effect) p_N will turn out to be a product of two primes, and now the needle is harder to find. In this case, program $\mathcal{P}$ will have a relatively easy job of it: Given candidate k, $\mathcal{P}$ need only test to see whether or not k divides p_N. The shortcut program $\mathcal{P}'$, by contrast, has the duty to find the needle for this N—and it is hard to see how $\mathcal{P}'$ is going to do this other than by testing various k's to see whether they divide p_N. So, here is an example in which (at least, *sometimes*) there does not appear a viable shortcut over the trial-and-error method. So, is this a counterexample to the assertion? Probably not. The problem is that $\mathcal{P}$ must do more than merely check whether k divides p_N—it must also check whether or not p_N is a product of exactly two primes (to know whether or not $k_0 = 0$). The difficulty (for $\mathcal{P}$) of doing this is comparable to the difficulty P' experiences in finding the needle in this case.

Either the assertion above is true, or it is false. It would be of great help in thinking about this subject, in my opinion, if we knew which. Indeed, as far as I am aware, we do not even have a counterexample to the stronger assertion that results from replacing the inequality with $f'(N) \leq h(N)$.

Chapter 17

Quantum-Assisted Computing

The discussion of the previous two sections suggests that the use of quantum mechanics may indeed gain efficiency for certain computations. However, at least three issues remain. First, as discussed in Sect. 16.4, our ability to apply quantum mechanics to specific problems appears to depend on finding a suitable technique for converting conventional computer programs to unitary operators. By "suitable," we mean a technique that results in no substantial loss of efficiency and is such that no entanglements are created with any scratch Hilbert spaces that must be introduced. Second, we must make allowance for the fact that, although programs act on arbitrary strings (and thus are suitable for computing real problems), our unitary operators always act on *finite* tensor products of H's. And, finally, we must find a suitable definition of "difficulty" for unitary operators. We now introduce a general framework for computations using quantum mechanics, a scheme that, among other things, addresses these three issues.

Fix, once and for all, the following objects: (i) a finite-dimensional Hilbert space H, (ii) a unit vector $|\psi_0\rangle$ in H, (iii) a finite list of unitary operators, each of which acts on some finite tensor product, $H \otimes \cdots \otimes H$, of H's, and (iv) a finite list of projection operators,[16] each of which acts on some finite tensor product of H's. The individual unitary and projection operators in these lists may operate on tensor products with different numbers of H-factors; for example, some may act on a single H, some on $H \otimes H$, etc. We label each unitary operator of (ii) and each projection operator of (iii) by a nonempty string (e.g., as U_S and $P_{S'}$, respectively); and, for later convenience, we do not use the same string to label both a unitary and a projection operator. (We shall later impose a further condition on this arrangement, but for the moment it is convenient to keep things general.)

We now introduce some terminology. First, we introduce a separator character $*$, in the manner we have done occasionally before. Next, we call a string $\tilde{S}$ a *unitary operation* if it is of the form $\check{S} * S_1 * \cdots * S_k$, that is, consists of $(k+1)$ strings (each nonempty and containing no $*$, and with $S_1 \cdots S_k$ distinct), such that: The first of these strings, $\check{S}$, labels some unitary operator $U_{\check{S}}$ in our list and that $U_{\check{S}}$ acts on a tensor product of precisely k factors of H. Thus, beyond the first string, it is only the *number* of additional strings, and not what those strings are, that counts. For example, if, among the unitary operators, there is one labeled U_{8k}, and if it acts on $H \otimes H$, then "$8k * yzr * 8\$9Q$" would be a unitary operation, whereas "$8k * yzr$" would not. And, similarly, we call string $\tilde{S}$ a *projection operation* if it is again of the form $\tilde{S} = \check{S} * S_1 * \cdots * S_k$ such that $P_{\check{S}}$ appears on our list of projection operators and it acts on the tensor product of exactly k factors of H. Note that no string is both a unitary operation and a projection operation; and that the problem of deciding whether a string $\tilde{S}$ is a unitary operation, a projection operation, or neither is computable and has difficulty function $L(\tilde{S})$.

We now introduce a new computer language. As before, we have storage locations, each of which is labeled by a

[16] A *projection* operator $\mathcal{P}$ is a self-adjoint operator satisfying $P \circ P = P$, that is, a self-adjoint operator having no eigenvalues other than 0 and 1.

string and each of which contains a string [where, as before, $C(S)$ denotes the string contained in location S]. There are seven commands in this language, consisting of the five we introduced in Sect. 12—INPUT, OUTPUT, APPEND, DELETE, and IF—together with two new ones:

6. APPLY $C(S)$
7. OBSERVE $C(S)$, APPEND RESULT TO $C(S')$

where, as before, S and S' denote arbitrary strings. Here is what these commands "do." In addition to the storage locations, there will now be a separate quantum system. The Hilbert space $\mathcal{H}$ of states of this system will be, at any one moment during the operation of the computer, some tensor product of H's, where each factor of H in this tensor product is labeled by a string. Thus, we might have, at one moment, $\mathcal{H} = H_8 \otimes H_{abc} \otimes H_{Q3}$, a tensor product of three copies of H. The state of this quantum system, at that moment, will be given by some vector, say $|\Psi\rangle$, in the Hilbert space $\mathcal{H}$. Now, here is what is to be done in response to the command APPLY $C(S)$:

1. If $C(S)$ is not a unitary operation, then do nothing.

2. If $C(S)$ $(= \check{S} * S_1 * \cdots * S_k$, say) *is* a unitary operation, and each of the strings $S_1, \cdots, S_k$ is already represented by an H-factor in the tensor product that is $\mathcal{H}$, then apply to the state $|\Psi\rangle \in \mathcal{H}$ the unitary operator $U_{\check{S}}$ on $H_{S_1} \otimes \cdots \otimes H_{S_k}$. (That is, the unitary operator that is applied to $\mathcal{H}$ is the operator $U_{\check{S}}$ applied to the factors $H_{S_1} \otimes \cdots \otimes H_{S_k}$, and I applied to the remaining factors.)

3. If $C(S)$ $(= \check{S} * S_1 * \cdots * S_k$, say) is a unitary operation, and some of the strings $S_1, \cdots, S_k$ are not represented by H-factors in the tensor product that is $\mathcal{H}$, then proceed as follows: First, enlarge $\mathcal{H}$ to include those H-factors (i.e., replace $\mathcal{H}$ by its tensor product with the missing H_S). Next, replace the state $|\Psi\rangle$ by the result

of taking the tensor product of this state with one copy of $|\psi_0\rangle$ for each new H-factor introduced. Finally, apply $U_{\check{S}}$ to this state in $\mathcal{H}$ (so enlarged) as in instruction 2.

Here is an example of these rules. Let, at some moment, $\mathcal{H} = H_{19} \otimes H_{yzr}$, let the state be $|\Psi\rangle \in \mathcal{H}$, let our list of unitary operators include an operator U_{8k} that acts on $H \otimes H$, and let $C(S) =$ "$8k * yzr * Q9$". Then APPLY $C(S)$ would replace this $\mathcal{H}$ by $H_{19} \otimes H_{yzr} \otimes H_{Q9}$ and state $|\Psi\rangle$ by $|\Psi\rangle|\psi_0\rangle$; and would then apply $I \otimes U_{8k}$ to this state.

Similar rules apply to OBSERVE $C(S)$, APPEND RESULT TO $C(S')$. If $C(S)$ is not a projection operation, do nothing. Otherwise, proceed as follows: First, enlarge $\mathcal{H}$ by including, as necessary, additional H-factors, labeled by those strings in $C(S)$ not already so represented. Next, replace the state $|\Psi\rangle$ by that state in this enlarged Hilbert space obtained by taking one tensor product with a $|\psi_0\rangle$ for each new factor of H. Then, make an observation on this new state of this expanded Hilbert space, via the self-adjoint operator $P_{\check{S}}$. The result of this observation must be either 0 or 1, since $P_{\check{S}}$ is a projection. Append this result (suitably encoded, if necessary) to the string in location $C(S')$. [Usually, we would have previously SET $C(S') = \emptyset$, to avoid clutter.] After this OBSERVation, the state of our quantum system will, of course, be replaced by its projection into the appropriate eigenspace, that is, by either $P_{\check{S}}$ or $(I - P_{\check{S}})$ applied to that state, according to whether the observation resulted in 1 or 0, respectively. It is generally more convenient to enlarge $\mathcal{H}$ by using the APPLY command, rather than the OBSERVE command.

In physical terms, what we are doing here is quite simple. We have some basic quantum system, described by Hilbert space H. What we call a projection operation, for example, is a string that describes a certain projection operator, and to which combination of copies of the basic system it is to be applied. If any of the required copies are missing, we simply supply them, by ordering new copies of that system (which come with state $|\psi_0\rangle$ preinstalled) through the catalog, and placing those new system-copies next to the old system-copies. When all the necessary copies of our basic system have been assembled, we observe via the appropriate (projection) operator and append the result to location S'. We remark that no generality has been lost by our making all observations via projection operators (rather than the more general self-adjoint operators). This is a consequence of the following fact: Every self-adjoint operator A on a finite-dimensional Hilbert space can be written as a sum, $a_1 P_1 + \cdots + a_s P_s$, where the a_i are the eigenvalues of A and the P_i project into the corresponding eigenspaces (so, in particular, the various P_i commute with each other). By virtue of this fact, we may, instead of observing via self-adjoint A, observe via each of the P_i, noting that, by commutativity, the order of the latter observations is irrelevant. These two operations will always produce the same result, in terms of the outcomes and their probabilities as well as the final state of the system.

A *quantum-assisted program* is a finite, ordered list of commands, the first command of which is INPUT and the last of which is OUTPUT, such that each of these two commands appears nowhere else in the program. We run a program just as before, starting with each storage location containing the empty string, and with $\mathcal{H}$ the (one-dimensional Hilbert space of) complexes. We then execute the commands in order, except as directed by IF. The various commands will then manipulate strings, or operate on, expand, or observe $\mathcal{H}$. So, for example, the first APPLY or OBSERVE command will require that $\mathcal{H}$ be expanded to include the appropriate copies of H as a tensor product. If and when the program reaches OUTPUT, it halts, allowing us to read the output string.

When a given quantum-assisted program is run with a given input string, it must either halt, with some output string returned, or never halt (which we denote, as before, by $*$). From the laws of quantum mechanics, there will be probabilities for these various outcomes; that is, we will have a probability distribution, p, on $\mathcal{S} \cup \{*\}$. We say that a quantum-assisted program *computes* problem π if, for every input string S, the program, run on that string, has $p(*) = 0$, and $p(\pi(S)) > p(S')$ for every $S' \neq S$. That is, the probability of not halting must be zero (although, as we have already seen, it may still be possible that the program fail to halt), and the probability of the "correct answer," $\pi(S)$, must exceed that of every other possible output.

As an example, let us consider the Grover construction. Let H be the two-dimensional Hilbert space of spin states of an electron, and let $|\psi_0\rangle$ be the state we earlier designated $|1\rangle$. Let U_1 act on H by $U_1|1\rangle = \frac{1}{\sqrt{2}}\{|1\rangle + |0\rangle\}$, $U_1|0\rangle = \frac{1}{\sqrt{2}}\{|1\rangle - |0\rangle\}$; U_2 by $U_2|1\rangle = -|1\rangle$, $U_2|0\rangle = |0\rangle$; and U_3 by $U_3|1\rangle = i|1\rangle$, $U_3|0\rangle = |0\rangle$. (A few more U's on H might also be required.) Let U_4 act on $H \otimes H$ by $U_4 = \exp[i(\pi/2)\vec{\sigma_1} \cdot \vec{\sigma_2})]$. Finally, let there be a single projection operator, P_5, acting on a single H via $P_5|1\rangle = |1\rangle$, $P_5|0\rangle = 0$.

Our program will accept as input a positive integer n. Let us now introduce three subroutines. The first places, in some location S, the string "$1*1$", then executes APPLY $C(S)$, then places "$1*2$" in location S, then executes APPLY $C(S)$ again, and so on up to a total of n times. The result of running this subroutine is to set up our Hilbert space, $H_{\text{in}} = H_1 \otimes \cdots \otimes H_n$, with each H_i-state given by $\frac{1}{\sqrt{2}}\{|1\rangle + |0\rangle\}$. (Note how we set up initial states, by starting with $|\psi_0\rangle = |1\rangle$ and applying the appropriate operator.) This is the setup subroutine: It arrange the initial configuration. The next subroutine applies to this H_{in} the operator V_{in} followed by the operator $-W$. We are assuming that we have already been given the subroutine for V_{in}, that is, that we have been given some rendering of our original needle-check program as a quantum-assisted program. The application of V_{in} may require an auxiliary Hilbert space, for scratch work, but we are demanding that in the final state the H_{in} and the scratch-work states not be entangled. Our earlier discussion of the construction of the operator W translates immediately into a quantum-assisted program, in our language, that applies this operator to H_{in}. Finally the third subroutine observes H_{in} via the k-basis. This will, for example, set $C(S) = 5 * 1$, execute OBSERVE $C(S)$, APPEND RESULT TO $C(S')$, and multiply the result [in $C(S')$] by 2^{n-1}; then set $C(S) = 5*2$, execute OBSERVE $C(S)$, APPEND RESULT TO $C(S')$, multiply the result [in $C(S')$] by 2^{n-2}, and add to the previous result; and so on, for n times around. This subroutine, then, will end up producing some k-value, where $0 \leq k \leq N-1$, stored in some memory location.

Now, there are at least three different quantum-assisted programs that could be constructed from these subroutines. For the first, run the setup subroutine, then run the subroutine $-WV_{\text{in}}$ once, then run the observe subroutine, and finally report the k-value that results. For this program, we have $p(*) = 0$ (and, in fact, outcome $*$ is impossible), $p(k_0) = 9/N$ (for large N), with the probabilities for the other k-values correspondingly reduced. This quantum-assisted program indeed computes our problem. For the second program, run the setup subroutine, then the subroutine $-WV_{\text{in}}$ the correct number [approximately $(\pi/4)\sqrt{N}$] of times, then the observe subroutine, and finally report the k value that results. For this program, we would again have $p(*) = 0$ (and, again, outcome $*$ is impossible). Now, however, $p(k_0)$ is nearly one [for $p(k_0) > 1 - 1/N$], whereas the remaining k's have probabilities nearly zero. This quantum-assisted program also computes our problem. For our final program, we first do the computation of the previous program, but instead of reporting the k that results from the observation subroutine, we instead run the (nonquantum) needle-check program on that k, to find out whether it is indeed the needle. If it is, we report that k-value. Otherwise, go through the entire procedure again, finding a new k and checking that one. Continue in this way until we find the k that checks out as the needle. Note that, for this program, we must assemble a fresh copy of H_{in} for each running of the prior program, for that program results in a final H_{in}-state different from the initial state needed by that program. For this program, we again have $p(*) = 0$, but now the outcome $*$ is possible (for we could to go on indefinitely, being unlucky time after time). We also have $p(k_0) = 1$; and $p(k) = 0$ for all other k. Thus, this quantum-assisted program also computes our problem. Thus, we have three separate quantum-assisted programs, each of which computes the problem (whatever it is).

In these examples, the OBSERVE commands generally come after the APPLY commands have already been executed. But that, of course, need not be the case in general: These commands could very well be intermingled. Note that the brain of this program is the original five commands and the storage locations with which they interact: These keep track of what is going on, decide when APPLY and OBSERVE are to be carried out, decide what to do with the results, handle the INPUT and OUTPUT, etc. The Hilbert space of the quantum system serves as a glorified storage register, into which we may place data (via APPLY), within which we may manipulate data (via APPLY), and from which we may extract data (via OBSERVE). We could also arrange, in effect, for there to be several different Hilbert spaces to handle data. These would be represented by different parts of the tensor product that is $\mathcal{H}$, and we would simply manipulate and access whatever part we wish to use at any one time, ignoring (i.e., applying the identity to, and not observing) the other parts. At any one moment in the program, the Hilbert space of states of the quantum system, $\mathcal{H}$, is a finite tensor product of H's, although, of course, the number of H's in this product is not limited. Had we, for example, replaced each $C(S)$ in the APPLY and OBSERVE commands with S, then there would, for any given program, be a limit (independent of the input string) on this number. Replacing each $C(S)$ in these commands by $C(C(S))$ would not make any difference, since we have a subroutine SET $C(S') = C(C(S))$. Note that the structure of quantum-assisted programs does not require, necessarily, that entanglements be avoided or that any entanglements that have been created eventually be undone. We simply APPLY certain unitary operators and OBSERVE certain projection operators. Whatever follows follows, whether there is entanglement or not. (Of course, it may be necessary to avoid entanglement in order that the program compute what we want it to compute.)

One point about this language should be emphasized. The quantum elements—the Hilbert space H, the initial state $|\psi_0\rangle$, the unitary operators, and the projection operators—are to be specified in full right at the beginning (once we know what problem it is we are to compute). It is not permitted to change these objects, depending on the input string.

Chapter 18

Quantum-Assisted Computability

We introduced, in the previous section, the notion of a problem being computable by a quantum-assisted program; and we already have, from Sect. 6, the notion of a problem being computable by a regular program. Clearly, every regular-computable problem is quantum-assisted-computable (since every regular program *is* already a quantum-assisted program, just one happening to have no APPLY and no OBSERVE commands). Is the converse true? That is, is it true that every problem that is computable with quantum assist is also computable in the regular sense?

This converse, as we have stated it above, turns out to be false. Here is an example. Let the Hilbert space H be two dimensional, with basis $|\psi_0\rangle$ (the initial state) and $|\psi_0\rangle^\perp$. Let there be just one unitary operator in the list, acting on a single copy of H by $U_\theta|\psi\rangle = |\psi_0\rangle\ \cos\theta + |\psi_0\rangle^\perp \sin\theta$, $U_\theta|\psi_0\rangle^\perp = |\psi_0\rangle^\perp \cos\theta - |\psi_0\rangle\ \sin\theta$, where θ is some fixed number. Let there be just one projection operator in the list, also acting on a single copy of H, by $P|\psi_0\rangle = |\psi_0\rangle, P|\psi_0\rangle^\perp = 0$. (You will note that this is a pretty poor excuse for quantum assistance: There are no operators in our lists that act on two or more H's, and so we will never produce, by means of these operators, entangled tensor-product states.) The program accepts as input a positive integer n and proceeds as follows: It first executes APPLY U_θ to H_1, OBSERVE P on H_1, and accumulates the result (0 or 1) in $C(a)$. It then repeats this subroutine, with H_1 replaced by H_2, and so on, all the way up to H_n. At this point, $C(a)$ will contain an integer (the number of times 1 was returned by OBSERVE during all n runs of the subroutine). Finally, the program returns, via OUTPUT, the rational number $C(a)/n$. What is this program doing? Well, on each run through the subroutine, either 1 will be added to the integer in $C(a)$ (with probability $\cos^2\theta$) or it will not (with probability $\sin^2\theta$). Thus, what the program returns at the end is a Monte-Carlo estimate of the value of $\cos^2\theta$, based on these n runs. The fractional error in this estimate goes down, as the number n of runs increases, as $1/\sqrt{n}$.

What this program does, in other words, is compute the number $\cos^2\theta$, in the sense of Sect. 7. What we have shown, then, is that for any number θ we can write a quantum-assisted program that computes $\cos^2\theta$. Now choose for θ that value such that $\cos^2\theta = c$, where c denotes the noncomputable number given by Eq. (2) of Sect. 7. Thus, we have produced a quantum-assisted program that computes a number that is not computable with any regular program.

The discussion of the previous two paragraphs will fool nobody. It is absurd to take seriously a unitary operator U_θ that claims to carry out a rotation through a noncomputable angle: We would not expect to be able to buy and operate a machine that would apply any such U_θ to any real quantum system. Suppose, for example, that we acquire a machine that is capable of carrying out a rotation in the $|\psi_0\rangle$–$|\psi_0\rangle^\perp$ plane through any angle θ, where the value of θ is set by adjusting a knob. Then, as we fine-tune this machine, we shall (in order to know where to set the knob) be called upon to determine, more and more precisely, what the number θ is. That is, we shall have have to decide whether or not more and more complicated Turing machines will halt. One of these, for example, is the machine that searches for a counterexample to the Goldbach conjecture—and so, at this point, the proper adjustment of the knob will require that we settle this conjecture, one way or the other. Moreover, as soon as we are finished with this one, we will be asked to resolve some other, even harder, conjecture in mathematics. How, in this atmosphere, are ever going to get this experiment finished? It is silly to call this quandary a "piece of laboratory equipment." Note that the fact that there are rational numbers arbitrarily close to the noncomputable number c is of no help to us here. The issue is one of *determining* what c is (in practice, in the laboratory), not one of *approximating* c (in principle, in mathematics land). To know c within 1%, let us say, will tell us whether or not the Goldbach conjecture is true. How helpful, in this circumstance, is it to be reassured that there does indeed exist a rational number within one-tenth of 1% of c?

> **Exercise.** Consider the unitary operator U_θ just described, but let us select the angle θ "randomly, by spinning, and then stopping the knob." Then the probability is 1 that we shall end up with a noncomputable number (since the computable numbers in $[0, 2\pi]$ are measurable, with measure zero). So here is an example in which a quantum-assisted computer computes a (regularly) noncomputable number. Respond.

So, we might ask, which unitary operators are, and which are not, physically realistic? In fact, the problem goes a little deeper than even this question suggests. Consider, for example, two unitary operators, U and U', each of which carries out a 90° rotation in the (three-dimensional, say) Hilbert space H, but such that the planes in which these rotations take place make angle c with each other. Although each of these two unitary operators, by itself, is quite innocent-looking, together they allow, in the same manner as before, a computation of c. The lesson here is that we cannot look at the state $|\psi_0\rangle$, the unitary operators that appear on our list, and the projection operators that appear on our list in isolation. It is the *entire list*—consisting of $|\psi_0\rangle$, the U's, and the P's; all taken together that must rise or fall. So, we might ask, how do we identify which such lists are, and which are not, physically realistic?

Here is a possible answer to this question. Let us imagine that the box in which the Hilbert space H is shipped has printed on it a standard basis for this Hilbert space, to be used for reference purposes. Then from this basis we acquire a standard basis for each tensor product, $H \otimes \cdots \otimes H$. Now, before purchasing a unitary operator U (on some tensor product of H's), we want to know what it is we are buying. This is to be provided by the manufacturer, in the form of a program, printed in the owner's manual, that accepts as input any positive integer n and returns rational numbers each within $1/n$ of the respective matrix element of U in this standard basis. We call this a program that *computes* U. Without such a program, we simply do not know what U is. We have, similarly, the notion of a program that *computes* a projection operator (on a tensor product of H's) and of a program that *computes* the initial state $|\psi_0\rangle$. We now regard H, state $|\psi_0\rangle$, and the lists of unitary and projection operators as "physically realistic" if, for some H basis, there exist programs that compute all these objects. This appears to be a rather mild condition.

Some terminology will allow us to formulate the idea of the previous paragraph more neatly. Fix, as before, (i) a finite-dimensional Hilbert space H, (ii) a unit vector $|\psi_0\rangle$ therein, (iii) a finite list of unitary operators (labeled by strings) on various H-tensor products, and (iv) a finite list of projection operators (labeled by strings) on various H-tensor products. We call a string $\mathbf{S}$ a *history* if it is of the form $S_1 * *S_2 * * \cdots * *S_k$, where each of the strings $S_1, \cdots, S_k$ is either (i) a unitary operation (string) or (ii) a projection operation (string) to which either $*0$ or $*1$

has been appended. Thus, a typical history string might be "$k9*yzr*0**B*xx*ABC$": The rightmost entry represents the unitary operator U_B applied to $H_{xx} \otimes H_{ABC}$; the other entry the projection operator P_{k9} applied to H_{yzr}, with $*0$ appended.

Now consider the running of some quantum-assisted program. Each time an APPLY command is executed, some things will be done with respect to the Hilbert space $\mathcal{H}$: This Hilbert space will be expanded (if necessary) by taking additional tensor products with H-copies; the state $|\Psi\rangle$ will be adjusted (if necessary) with $\otimes|\psi_0\rangle$'s to lie in this expanded Hilbert space; and a certain unitary operator will be applied to this state. The same goes for an OBSERVE command, except that in this case either the projection operator P (from the list) is applied to the state (the case in which the observation yielded 1) or the projection operator $(I - P)$ is applied (if the observation yielded 0). The other commands, INPUT, OUTPUT, DELETE, APPEND, and IF, do nothing with respect to $\mathcal{H}$. Thus, as of any one moment during the running of the program, $\mathcal{H}$ will have been subjected to some finite number of such operations, in some order. But this is precisely the information contained in a history. In other words, a history string provides a complete summary of what has happened with respect to $\mathcal{H}$ as of a certain moment. Perhaps "virtual history" would be a better term, for we admit as histories *all* strings formed by the rules of the previous paragraph, whether or not they happen to represent what has actually occurred. From the history string we may determine what the Hilbert space $\mathcal{H}$ is at that moment, what state (in $\mathcal{H}$) the quantum system is in, and what information (0's and 1's) has been passed so far from the quantum system to storage locations. From the history $\mathbf{S}$ = "$k9*yzr*0**B*xx*ABC$" of the previous paragraph, for example, we would determine that $\mathcal{H} = H_{yzr} \otimes H_{xx} \otimes H_{ABC}$, that the state is $((I - P_{k9}) \otimes I \otimes I)((I \otimes U_B)(|\psi_0\rangle|\psi_0\rangle|\psi_0\rangle))$, and that value 0 was returned on the execution of the one

OBSERVE command. The idea, in short, is to reflect the entirety of the quantum part of quantum-assisted computing by a string, something we can easily dissect.

For $\mathbf{S}$ any history, denote by $\gamma(\mathbf{S})$ the squared norm of the state (as determined by $\mathbf{S}$) in the Hilbert space (as determined by $\mathbf{S}$). Thus, for example, we have $\gamma(\mathbf{S}) = 1$ if the history $\mathbf{S}$ contains no projection operations (since $|\psi_0\rangle$ is unit, and unitary operators are norm-preserving); for a general history $\mathbf{S}$, $0 \leq \gamma(\mathbf{S}) \leq 1$. This real-valued function γ on histories carries all the relevant information about the workings of quantum mechanics within our quantum-assisted language, in a sense that will become clear shortly.

Now suppose that, with respect to some standard basis for H, there exists a program that computes the initial state $|\psi_0\rangle$, as well as ones that compute each of the unitary and projection operators in our lists, in the sense just described. Then we may combine these to produce a program that, given any history $\mathbf{S}$, will compute the components, in terms of this standard basis, of the state determined by this $\mathbf{S}$. It is now a simple matter to write a program that computes the function γ, in the following sense: This program accepts as input any history $\mathbf{S}$ and positive integer n and returns some rational number within $1/n$ of $\gamma(\mathbf{S})$. The existence of such a program implies, of course, that each number $\gamma(\mathbf{S})$ is computable, but it also implies a great deal more: It means that there is a *single* rule that suffices to provide an approximation for *every* $\gamma(\mathbf{S})$.

We now return to the issue raised at the beginning of this section. We claim: Any problem that is computable by a quantum-assisted program, using computable initial state and operators, is also computable by a regular program. The proof, much like that of the similar result for probabilistic programs, is by simulation.

Fix H, and let $|\psi_0\rangle$ and the labeled unitary and projection operators be computable, in the sense above. Now fix a quantum-assisted program $\mathcal{P}_{\text{quant}}$, together with the input string, S_{in}, on which this program is to be run. We are going to construct a regular program, $\mathcal{P}_{\text{reg}}$, that will simulate the running of $\mathcal{P}_{\text{quant}}$ on S_{in}. Suppose that $\mathcal{P}_{\text{quant}}$ has been run for a few steps, encountering an APPLY command or two, but no OBSERVE command. Then the entire state of the computer (including the quantum system) can be expressed, as of this moment, by giving three pieces of information: (i) which command in the list $\mathcal{P}_{\text{quant}}$ is slated to be executed next, (ii) which string is stored in each (nonempty) storage location, and (iii) the history, **S**, of the quantum system. Let us now carry $\mathcal{P}_{\text{quant}}$ through the next step (say, a non-OBSERVE one). To simulate this step, we update the three pieces of information in the obvious way: For (i), we now indicate what is the new next command; for (ii) we make an adjustment (required only if that step was APPEND or DELETE, and then to only one of the stored strings) to reflect the new string storage; and for (iii) we add an entry (required only if that step was APPLY) to the history. In this way, we continue our simulation of $\mathcal{P}_{\text{quant}}$, step by (non-OBSERVE) step. What happens when we reach an OBSERVE command? Now there will be two possible outcomes, depending on whether the observation returns 1 or 0. We reflect this state of affairs by splitting our description into two branches, each of which carries three pieces of information as before. In one branch, corresponding to the observation returning 1, the three pieces of information read as follows: for (i), the next command to be executed; for (ii), the same stored strings, but with 1 appended to one particular string; and, for (iii), the same history string, but with the addition of a certain projection operation and $*1$. In the other branch, we enter, similarly, the three pieces of information appropriate to the case in which the observation returns 0. We now continue to simulate the behavior of $\mathcal{P}_{\text{quant}}$ in each of these two branches separately. As more OBSERVE commands are executed, the number of branches will grow, as will the burden of separately simulating what happens within each branch. But at every stage in this simulation, we shall have a finite number of branches, each described by just these three pieces of information.

So, $\mathcal{P}_{\text{reg}}$ simulates $\mathcal{P}_{\text{quant}}$, in this way. Every so often, one of the branches being simulated by $\mathcal{P}_{\text{reg}}$ will reach an OUTPUT command (after which there is nothing more to simulate). When that occurs, the program $\mathcal{P}_{\text{reg}}$ reads the output string S and the final history $\mathbf{S}$ for that branch, stores this information in a special section, and drops that branch from further consideration. [We, of course, know that $\gamma(\mathbf{S})$ is the probability that the actual program $\mathcal{P}_{\text{quant}}$ will take this branch, reaching this OUTPUT and returning this S.] Thus, as the simulation by $\mathcal{P}_{\text{reg}}$ continues, the list of string-history pairs in this special section will grow. Our program $\mathcal{P}_{\text{reg}}$ will include, furthermore, a subroutine, which accepts as input a positive integer n, and operates as follows: The subroutine goes to this special section, takes each of the history strings in that section, and computes γ of that history, within error $1/n$ (here using the program that we constructed from the reference manuals). It then totals these numbers, for each output string listed in that section. Finally, the subroutine checks to see whether any one output string, say S, has emerged as a clear winner (i.e., is such that no other string S' will ever be able to achieve a total exceeding that for S, even if we allocate to S' all the so-far unallocated probability, and even if we assume that all the $1/n$-errors in these probabilities are resolved in S''s favor). If the subroutine finds such a clear winner S, then it causes $\mathcal{P}_{\text{reg}}$ to halt immediately, giving that S as output. If there is no clear winner, then the subroutine returns $\mathcal{P}_{\text{reg}}$ to its simulation.

The full program $\mathcal{P}_{\text{reg}}$ now operates as follows: Every so often (say, after every 100 steps of simulation), $\mathcal{P}_{\text{reg}}$ runs the subroutine, using an n-value one higher than that for the previous run. So, this program $\mathcal{P}_{\text{reg}}$ will continue to run in this way: continuing to carry out the simulation of $\mathcal{P}_{\text{quant}}$, continuing to store the results for halted branches in the special section, and continuing to make ever-finer checks on the status of that special section. Now suppose that the program $\mathcal{P}_{\text{quant}}$ computes a problem π, that is, that, for every input string S_{in}, $\mathcal{P}_{\text{quant}}$ has probability zero of failing to halt and has probability of halting with output $\pi(S_{\text{in}})$ that exceeds that of every other possible output. In this case, our simulation $\mathcal{P}_{\text{reg}}$ must eventually halt, for eventually it will have accounted for sufficient probability to identify $\pi(S_{\text{in}})$ as the clear winner. At this point, $\mathcal{P}_{\text{reg}}$ will return $\pi(S_{\text{in}})$.

What we have shown, then, is that, given any quantum-assisted program that computes a problem, we can, by simulating it in this manner, build a regular program that computes the same problem. Note that $\mathcal{P}_{\text{reg}}$ *always* halts, giving the correct answer, $\pi(S_{\text{in}})$. The quantum-assisted program $\mathcal{P}_{\text{quant}}$, by contrast, gives this answer only probabilistically. Note also that this argument makes essential use of the program that computes the function γ. In any case, we conclude that any problem that is computable by a quantum-assisted program, with computable $|\psi_0\rangle$ and operators, is also computable by a regular program.

The result of this section is of perhaps only mild interest. Of greater importance are the notions of a history, of the function γ, and of the present strategy for simulating a quantum-assisted computation. We shall use all of these extensively in what follows.

Chapter 19

Quantum-Assisted Difficulty Functions

In Sect. 10, we defined a difficulty function for every program that computes some problem π. This positive, real-valued function represents the amount of "computer time," as a function of the input string S_{in}, required to compute $\pi(S_{\text{in}})$. We now wish to do the same thing for any quantum-assisted program that computes a problem. We shall do so in two steps. First, we assign a difficulty to each individual command in our quantum-assisted language. This will entail a certain restriction on the unitary and projection operators that go into the language. Second, we adapt the notion of a difficulty function to take into account the fact that quantum-assisted computing provides answers only probabilistically.

We emphasize again that we have set things up so that the quantum elements—the Hilbert space H, the initial state $|\psi_0\rangle$, and the unitary and projection operators—are to be specified as part of the program; that is, they, may depend on the problem to be computed, but not on the particular input string. Imagine that one had done otherwise—for example, had allowed the use of different unitary operators for different input strings. Then how would we produce any reasonable notion of a difficulty function? It could be, for example, that certain input strings would require unitary operators that are very delicate—ones that can be created only with a great deal of time and effort. But this "time and effort" will have to go into the difficulty for those input strings. That is, to produce a difficulty function in such a language we would have to quantify the delicacy of unitary operators. Such a project—to put it mildly—does not look easy. We remark that many of the "quantum computations" that have been proposed suffer from precisely this defect.

There are seven types of commands that may appear in a quantum-assisted program. Five of these—INPUT, OUTPUT, APPEND, DELETE, and IF—are the original commands we introduced in Sect. 12, and it is natural to assign to these commands the difficulties we already chose earlier. But what of the two new commands—APPLY and OBSERVE? Note that we have finite lists of the unitary and projection operators that are permitted to appear in these commands. So, in effect, each of APPLY and OBSERVE represents a finite number of physical operations. A natural choice would thus seem to be the following: Assign, to each APPLY and OBSERVE command, difficulty one. However, unfortunately, things are not that simple, as the following example illustrates.

Consider a problem, π, that accepts as input a positive integer n, returning either 0 or 2, and is such that every program that computes this π has difficulty function f that grows very quickly with n (say, faster than 2 to the power of 2 to the power of 2, etc., for n iterations). We have seen in Sect. 11 that there does indeed exist such a (computable) problem. Now set $c' = \sum_n \pi(n)/3^n$, a certain real number. Thus, this c' is constructed in the same manner as the non-computable number c of Sect 6; but, in contrast to that c, this c' is a computable number (since the problem π is). The point, however, is that c' is hard to compute: The number of computer steps required to approximate c' within $1/n$ grows very quickly with n. Let us now consider a quantum-assisted language in which the Hilbert space H is two dimensional and the list of unitary operators includes a U_θ that applies a rotation, to a single H, through angle θ, just as in the previous section. Now, however, we choose $\cos^2\theta = c'$. We next write a quantum-assisted program in this language similar to that of the previous section. That is, this program repeatedly applies U_θ and makes an observation, resulting in a Monte-Carlo estimate of $\cos^2\theta$. To compute $\pi(n)$, we must estimate the value of c' to within $1/(2 * 3^n)$. But the error in a Monte Carlo estimate decreases as the reciprocal of the square root of the number of runs. Thus, we need about $(2 * 3^n)^2 = 4 * 9^n$ applications of U_θ to have a reasonable chance of recovering the value of $\pi(n)$. For given n, carry out 10^{n+1} applications (just to be on the safe side): Then we shall have a probability of correctly determining $\pi(n)$ that is high and increasing with n. Thus, we have written a quantum-assisted program that computes this problem π with, presumably, difficulty function 10^{n+1}—much less than the difficulty function of any regular program that computes π.

This was a foolish argument in the previous section, and it is no better this time around. What this argument does show, however, is that we must be prepared to exercise some care as to which unitary (and projection) operators will be allowed in quantum-assisted programs, and as to what their difficulties are to be.

It is tempting to take the position that, since this U_θ is apparently such a terribly difficult operator, we can make things right by merely assigning a large difficulty to the corresponding APPLY command. But this is not going to work: There is just one of these U_θ's, and there is just one corresponding APPLY command, and so there is just one number for us to assign. Changing the difficulty of this single command from 1 to 1,000, for example, will increase the difficulty function by at most a factor of 1,000, that is, will result in an equivalent difficulty function, and so will not undermine the above argument. In fact, it appears that any attempt to preserve this particular U_θ in our list of unitary operators will return us to the issue of how we take into account errors. What is problematic about this operator U_θ appears only in attempts to "approximate" it. Imagine a new kind of quantum-assisted program that, when commanded to APPLY this U_θ, actually applies an operator that is only a rough approximation to U_θ, but which (by virtue of the roughness of the approximation) is also low in difficulty. If and when, as the running of the program proceeds, a more accurate application of U_θ becomes necessary, then our program would go back and redo the original APPLY command, but this time applying something that is closer to the actual U_θ (and carries a larger difficulty).

So, it appears that the only way we can avoid a very complicated programming environment, in which errors must constantly be taken into account, is to banish this U_θ from out list of unitary operators. But this is a slippery slope: Will there be allowed other unitary operators, based on numbers that are a little easier to compute, but still pretty hard? Where do we draw the line? Our approach will be to go ahead and slide down the slope, that is, to rule out all but the "simplest" U's.

Fix an m-dimensional Hilbert space H, a unit vector $|\psi_0\rangle$ in H, and finite lists of unitary and of projection operators, each acting on some finite tensor product of H's. We say that this arrangement is *simple* if, for every history string $\mathbf{S}$, the number $\gamma(\mathbf{S})$ is rational, and, furthermore, there exists a (regular) program that computes this γ, with difficulty function f satisfying

$$f(\mathbf{S}) \leq N_{\text{op}}(\mathbf{S})\; m^{N_{\text{H}}(\mathbf{S})}. \tag{13}$$

Here, $N_{\text{op}}(\mathbf{S})$ denotes the total number of (unitary or projection) operations represented by the history string $\mathbf{S}$, N_{H} denotes the total number of copies of the Hilbert space H that appear in the final tensor product [within which $\gamma(\mathbf{S})$ is computed], and m is the dimension of H. Note that simplicity implies immediately that γ is computable, in the sense of the previous section.

Here is why this definition is what it is. Fix some basis for H, say $|\alpha_1\rangle, |\alpha_2\rangle, \cdots, |\alpha_m\rangle$. Then, as we have seen, we may construct from this H-basis a basis for every tensor product,

$H \otimes H \otimes \cdots \otimes H$, of H's. For n H's in the tensor product, this basis will contain m^n vectors, each of which is a product of some n vectors from our H-basis. Now suppose that, with respect to this basis, the components of $|\psi_0\rangle$ and of all the unitary and projection operators in our list are rational numbers. This is about as simple as $|\psi_0\rangle$ and the U's and P's could possibly be. Now, in this case each value of γ will certainly be rational. Furthermore, we can easily write a program that computes γ. This program would take the history **S** and express explicitly, in terms of our basis, the result of applying in succession each operation contained in **S**. The program then takes the resulting final state, again expressed in terms of components in this basis, and computes its squared norm.

What is the difficulty function of this program? Consider one of the operations—say, application of some unitary U—in the history string **S**, and suppose that, at the point at which this U is applied, the Hilbert space is a tensor product of n copies of H. Then, at that point, the dimension of this Hilbert space will be m^n, and so the current state will have m^n components, and so the record of this state in our program will require that m^n entries be stored. To apply the operator U to this state will entail replacing each of these entries by a linear combination of other entries. That is, to compute the effect of this U we shall have to carry out a number of arithmetic operations given by a small multiple of m^n. Note—and this is a key point—that there is no savings from the fact that U actually operates on a small number of H's in that tensor product. For example, say that H has basis $|0\rangle, |1\rangle$, and let U act on a single H, by $U|1\rangle = \frac{4}{5}|1\rangle + \frac{3}{5}|0\rangle$, $U|0\rangle = \frac{4}{5}|0\rangle - \frac{3}{5}|1\rangle$. Let the current tensor product consist of a large number n of H's, and suppose that this U is acting on the 73rd one. Consider any two basis elements,

$$|0\rangle|1\rangle|1\rangle|0\rangle \cdots |1\rangle \cdots |0\rangle|1\rangle,$$

$$|0\rangle|1\rangle|1\rangle|0\rangle \cdots |0\rangle \cdots |0\rangle|1\rangle,$$

differing only in their entry for the 73rd copy of H. Now, the action of U will mix these two elements. Thus, to compute how this U acts, we will have to carry out a small arithmetic computation involving the component-values stored in these two locations. But the same is true for all m^n component-values stored. So, the order of m^n arithmetic operations must be carried out. And, apparently, there is no available shortcut, by which multiple entries can be calculated or stored all in one shot. The next application of a U may involve the 194th H in the tensor product, and computing its action will again involve the entries in *all* the m^n locations, grouping those entries in a different way from that of the previous application of U.

So, under the assumption of rational components in a certain basis, the difficulty required to compute the effect of application of one U or P in our list is a small multiple of m^n. So, the total difficulty to compute the rational number $\gamma(\mathbf{S})$ is a sum of terms, one for each operation in the history **S** and each of the form m^n, where n is the then-number of H's in the Hilbert space. Inequality (13) is a simpler, and somewhat weaker, expression of this bound. We conclude: In the case in which $|\psi_0\rangle$ and the U's and P's have rational coefficients in some H-basis, that arrangement is simple as defined here. In fact, a few other cases are also allowed by the definition (e.g., that in which U is of the form $U|1\rangle = \frac{1}{\sqrt{2}}\{|1\rangle + |0\rangle\}$, $U|0\rangle = \frac{1}{\sqrt{2}}\{|1\rangle - |0\rangle\}$). The definition of "simple" as given has the advantages that it allows these other cases and also that it makes no reference to any basis.

So, in short, a system—of $|\psi_0\rangle$, some unitary U's, and some projection P's—is simple if the only thing that counts in computing $\gamma(\mathbf{S})$ is the number of operations represented by **S** and the size of the Hilbert spaces on which these operations act. There is no factor to represent "how hard" the arithmetic manipulations are. Simplicity means, in effect, that the operators require only "easy" arithmetic.

We are now in a position to appreciate the key difference between a quantum-assisted program and a regular program.

The quantum-assisted program can apply one of its unitary or projection operators in a single step. This is because the operators themselves are rather simple, and each of them applies to only a few H's. The quantum-assisted computer simply assembles the appropriate two or three H's and applies the operator—all without even knowing about any other H's that may be involved in the tensor product. But, in order for a regular program to see what is happening, it is necessary for that program to consider *all* the H's in the tensor product: It cannot simply ignore those H's to which the operator does not apply. In short, quantum mechanics is able to *do* (easily, and probabilistically) what nonquantum mechanics can only *compute* (with much more difficulty, and numerically). This state of affairs is reflected by the fact that the regular program ends up with a difficulty function for γ satisfying (13), whereas the analogous inequality for a quantum-assisted program would read $f_{\text{quant}}(\mathbf{S}) \leq N_{\text{ops}}(\mathbf{S})$. What quantum mechanics has going for it, in short, is the tensor product.

So, we have decided what combinations of H, $|\psi_0\rangle$, U's, and P's (namely, the simple ones) to allow in our quantum-assisted programs as well as what difficulty (namely, one each) to assign to the new commands, APPLY and OBSERVE, in that language. We must now contend with the probabilistic aspect of quantum-assisted computing.

Fix a quantum-assisted program, $\mathcal{P}_{\text{quant}}$, that computes some problem, π, in the sense of Sect. 17. Thus, for every input string, S_{in}, we have a probability distribution p on the possible outcomes with this, S_{in}, and these satisfy $p(*) = 0$ and $p[\pi(S_{\text{in}})] > p(S')$ for every $S' \neq \pi(S_{\text{in}})$. We wish to assign a difficulty function to this entire program. The situation here is similar to that we faced in our discussion, in Sect 13, of probabilistic computing. We must take account of the fact that different runs of our program may require different numbers of steps and also that the output from running our program, on input string S, may be something other than the right answer, $\pi(S)$. And we adopt the same formula as we obtained earlier: We assign, for the difficulty for the computation on input string S, the value $f(S) = D(S)(p+p')/(p-p')^2$, where $D(S)$ is the expected difficulty in running this program on this input string, p denotes the probability of the correct output, $\pi(S)$, and p' denotes the probability of the next most probable output. The first factor on the right corrects for the fact that different runs of our program may encounter different difficulties—we take the mean difficulty. The second factor corrects for the probabilistic nature of the output.

As an example of these ideas, consider again the Grover construction, as reflected by the three distinct quantum-assisted programs introduced in Sect. 17. We now determine the difficulty function for each of these programs. For input n any a positive integer, denote by $h(n)$ the difficulty encountered by a quantum-assisted subroutine in applying the entire operator WV to the tensor product, $H \otimes \cdots \otimes H$, of n H's, so $h(n) \geq n$. Then, as part of the lore of this construction, this same $h(n)$ will be the largest difficulty encountered by a regular program making a check to see whether a single k is the needle in the n-haystack. Thus, a regular program can compute this problem with difficulty function $f_{\text{reg}}(n) = Nh(n)$, where $N = 2^n$ is the total number of needles in the haystack.

In the first quantum-assisted program, a single iteration of WV is made, followed by a series of n OBSERVations. There results a candidate k for the needle, which is then immediately reported using OUTPUT. The probability (p) that this k is the actual needle is about $9/N$; while the remaining k-values share the rest of the probabilities (so each p' is about $1/N$). The mean difficulty in this case is $D(n) = h(n) + n$ (since a single iteration is performed followed by n observations). Substituting into our formula, we obtain a difficulty function (for large N) $f_{\text{quant}}(n) = (10/64)(h(n)+n)N$, which is equivalent to the difficulty function, obtained earlier, of the regular program. It should come as no surprise that *this* strategy for a quantum-assisted computation brings no advantage.

In the second program, a total of $\sqrt{N}$ (give or take a few) iterations of WV are made, again followed by a series of n OBSERVations, and the reporting of a needle candidate. Here, the mean difficulty is $D(n) = \sqrt{N}h(n) + n$. The probability that the reported k is the actual needle is now about $p = 1 - \frac{1}{N}$, while the remaining k-values share the rest of the probabilities (so p' is approximately $\frac{1}{N^2}$). Substituting into our formula, we obtain difficulty function $f_{\text{quant}}(n) = (\sqrt{N}h(n) + n)(1 - \frac{1}{N} + \frac{1}{N^2})/(1 - \frac{1}{N} - \frac{1}{N^2})^2$. This function, for large N, is equivalent to $\sqrt{N}h(n)$. Note that the difficulty function for this program is $\ll$ than the difficulty function for the regular program, reflecting a potential advantage for the quantum assist (which would, perhaps, be a real advantage, if only we had a good candidate for what problem is being computed here).

In the third program, we begin, just as before, with a total of $\sqrt{N}$ iterations of WV, followed by a series of n OBSERVation. But in this case we check, using the regular program, whether or not the k that results is indeed the needle. If it is, report that k [thus incurring total difficulty $\sqrt{N}h(n)+n+h(n)$], but if it is not, go back to the beginning, carrying out the iterations and the OBSERVations again. Repeat until you find the needle. In this case $p = 1, p' = 0$ (since we will either find the needle or—with probability zero—continue trying forever). But now the mean difficulty (which really is a mean in this case, for now there is a nontrivial probability distribution on difficulties) is more complicated. The probability that we carry out just one group of $\sqrt{N}$ iterations of WV is $1 - \frac{1}{N}$ (approximately), that we carry out two is $\frac{1}{N}(1 - \frac{1}{N})$, etc. So, the mean difficulty is given by

$$D(n) = [\sqrt{N}h(n) + n + h(n)]\left[1\left(1 - \frac{1}{N}\right) + 2\frac{1}{N}\left(1 - \frac{1}{N}\right) + 3\frac{1}{N^2}\left(1 - \frac{1}{N}\right) + \cdots\right]. \tag{14}$$

The sum on the right is $N/(N-1)$. Substituting these into our formula, we obtain, for large N and up to equivalence, the difficulty function of this program: $f_{\text{quant}}(n) = \sqrt{N}h(n)$. This is identical to the difficulty function of the previous program.

These are precisely the results that we expect. The first program is not really exploiting the potential advantages of quantum mechanics, and its difficulty function shows it. The last two are essentially the same program. The only difference is that the first program has a fixed difficulty per run (as opposed to a probability distribution in difficulties), but it leaves some unfinished business in the form of the output probabilities, whereas the second yields certainty for the correct output, at the cost of possibly requiring several repetitions. Our definition of the difficulty function of a quantum-assisted program is so constructed to ignore such window-dressing.

We remark that one way to exploit quantum mechanics in the computation process is to use it merely as a random-number generator. That is, there would not, in the course of the computation, be created any entanglements between the various H-factors in the tensor product. Instead, we would simply create an H-factor (using APPLY), with initial state $|\psi_0\rangle$, and then immediately thereafter make an observation on that factor (using OBSERVE), resulting in various probabilities for the various outcomes. These two steps would be repeated, as necessary, throughout the program. Indeed, every probabilistic program (as defined in Sect. 13) can in this manner be simulated by a quantum-assisted program. For the quantum-assisted programs in this class, each $\gamma(\mathbf{S})$ that will have to be evaluated can be computed with (up to equivalence) difficulty one (as follows from the fact that no entanglements are created between the H-factors). Note further that the difficulty function for a quantum-assisted program is virtually a rewrite of the difficulty function for a probabilistic program. There follows from these remarks:

Theorem. Let π be a problem, and let $\mathcal{P}_{\text{prob}}$ be any probabilistic program that computes that problem. Then there exists a quantum-assisted program $\mathcal{P}_{\text{quant}}$ that also computes π, such that $\mathcal{P}_{\text{quant}}$ has precisely the same difficulty function as $\mathcal{P}_{\text{prob}}$.

In other words, any benefits that probability might bring to the computation process are also borne by quantum mechanics. In Sect. 13, we remarked that it is, apparently, an open question whether there exists a problem such that some probabilistic program computes that problem more efficiently than any regular program. If there were such a problem, then it follows from the theorem that some quantum-assisted program would also compute that problem more efficiently than any regular program. Of course, the status of the converse of this assertion is not clear: It could turn out that there exists no problem for which probability increases efficiency, and yet there does exist a problem for which quantum mechanics increases efficiency.

Chapter 20

Quantum-Assisted Efficiency I

This completes our formulation of quantum-assisted computing. This formulation begins by fixing a character set, together with a finite-dimensional Hilbert space H, a state in that Hilbert space, and finite lists of unitary and projection operators, each acting on some finite tensor product of H's. On these objects we impose the condition of simplicity. We then introduce a quantum-assisted programming language, consisting of some seven commands. We introduce the notion of a program's computing a problem, as well as the difficulty function associated with such a program. These are the building blocks of quantum-assisted computing. In this section and the next, we compare quantum-assisted programs and regular programs with respect to their difficulty functions. Here, we obtain a result to the effect that the maximum reduction in difficulty that can be achieved by quantum assist is logarithmic.

Fix a quantum-assisted program, $\mathcal{P}_{\text{quant}}$, that computes some problem π, and denote its difficulty function by f_{quant}. We construct a regular program, $\mathcal{P}_{\text{reg}}$, that simulates $\mathcal{P}_{\text{quant}}$, in the following manner: Fix the input string, S_{in}. Then $\mathcal{P}_{\text{reg}}$ simulates the running of $\mathcal{P}_{\text{quant}}$, on this input string, in the same manner as in Sect. 18. That is, at any one moment $\mathcal{P}_{\text{reg}}$ is following a number of "branches" of $\mathcal{P}_{\text{quant}}$ (each spawned by the simulated execution of an OBSERVE command); and for each of these branches $\mathcal{P}_{\text{reg}}$ keeps track of three pieces of information: (i) what is the next command, in the list $\mathcal{P}_{\text{quant}}$, to be executed; (ii) what is stored, by $\mathcal{P}_{\text{quant}}$, in all nonempty storage locations; and (iii) what is the history string $\mathbf{S}$ representing interactions $\mathcal{P}_{\text{quant}}$ has initiated with the quantum system. We now modify that earlier simulation, in two ways.

First, the earlier simulation (implicitly) proceeded along each branch at the same command rate. That is, one additional command was executed in every branch, then one more command in every branch, and so on. Now, however, we proceed along each branch at the same difficulty rate. That is, we carry out one unit of $\mathcal{P}_{\text{quant}}$-difficulty in each branch, then one more unit in each branch, etc. Thus, branches that involve a great deal of difficulty per command are simulated more slowly than those that involve less.

For the second modification, recall that in the earlier simulation $\mathcal{P}_{\text{reg}}$ maintained in its memory a special section, which was added to each time a branch under simulation reached a "halt" (i.e., a $\mathcal{P}_{\text{quant}}$-OUTPUT command). When this occurred, the program $\mathcal{P}_{\text{reg}}$ stored in this section the current history string $\mathbf{S}$, as of that halt, as well as the string S that would have then been returned by $\mathcal{P}_{\text{quant}}$. A branch, once reported in this way was then abandoned by $\mathcal{P}_{\text{reg}}$. The present simulation is a little different. The special section now contains a certain list of strings and, for each such string S, a corresponding rational number. When a branch, while under simulation, reaches a halt, $\mathcal{P}_{\text{reg}}$ immediately computes the rational number $\gamma(\mathbf{S})$ (where $\mathbf{S}$ is the current history string), adds this number to the number already stored for string S (where S is the string that $\mathcal{P}_{\text{quant}}$ would have returned), and then again abandons that branch. Thus, the various strings listed in this special section are, as before, the possible outputs from $\mathcal{P}_{\text{quant}}$ up to this point. But now the (rational) number stored for each string gives the total probability that

$\mathcal{P}_{\text{quant}}$ would, by this point, have returned that string. In addition, $\mathcal{P}_{\text{reg}}$ contains a subroutine, which operates as follows: It goes through the list of strings and (rational) probabilities in the special section and determines whether any output string in that list can be declared a clear winner (i.e., has a total that is greater than that which could be achieved by any other string, even if that string were allocated all so-far unallocated probability). If the subroutine finds a clear winner, then $\mathcal{P}_{\text{reg}}$ itself halts, returning the winning string. This subroutine is run each time $\mathcal{P}_{\text{reg}}$ finds itself making an addition to the special section.

So, given the program $\mathcal{P}_{\text{quant}}$, we may write this program $\mathcal{P}_{\text{reg}}$, which, for every input string, simulates the behavior of $\mathcal{P}_{\text{quant}}$, as just described. Clearly, this $\mathcal{P}_{\text{reg}}$ always halts, and it computes the same problem as $\mathcal{P}_{\text{quant}}$ does. Denote by f_{reg} the difficulty function of $\mathcal{P}_{\text{reg}}$. The plan is to use (13) to find an inequality that bounds f_{reg} in terms of f_{quant}.

Fix the input string S_{in}, and denote by p the probability that $\mathcal{P}_{\text{quant}}$ will return $\pi(S_{\text{in}})$ and by p' the probability of the next-most-likely output, so $p > p'$. Then $\mathcal{P}_{\text{reg}}$ will be able to declare a clear winner, and so will halt, at least by the time it has accounted for an amount $1 - \frac{p-p'}{2}$ of probability.[17] Denote by $\mathcal{N}$ the total amount of $\mathcal{P}_{\text{quant}}$-difficulty that $\mathcal{P}_{\text{reg}}$ has simulated (in each branch) at the point at which $\mathcal{P}_{\text{reg}}$ halts. Then we have

$$f_{\text{quant}}(S_{\text{in}}) = D(S_{\text{in}})\frac{p+p'}{(p-p')^2} \geq \left\{\mathcal{N}\frac{p-p'}{2}\right\}\frac{p+p'}{(p-p')^2} \geq \frac{1}{2}\mathcal{N}. \tag{15}$$

The first step in (15) is the definition of f_{quant}. The second step uses the fact that $\mathcal{P}_{\text{reg}}$ has already gone through amount

[17] To see this, denote by x the amount of probability that $\mathcal{P}_{\text{reg}}$ has accounted for up to some point. In the worst-case scenario, an amount $1-p$ (the maximum possible) would have been used on the non-p outcomes (including amount p' on the p'-outcome), leaving just $x - 1 + p$ for the p-outcome. So, for there to be declared a clear winner at this point, p's amount $(x - 1 + p)$ must exceed p''s amount (p') plus the so-far unallocated probability $(1 - x)$.

$\mathcal{N}$ of $\mathcal{P}_{\text{quant}}$-difficulty, and yet there still remains probability at least $\frac{p-p'}{2}$ that $\mathcal{P}_{\text{quant}}$ has not halted. This fact alone contributes to the mean total difficulty of $\mathcal{P}_{\text{quant}}$ an amount equal to the product of these two numbers.

We must now relate f_{reg} to this $\mathcal{N}$. To this end, denote by M_{H} the maximum number of additional H's that can be introduced into the tensor product per unit difficulty. For example, if no unitary or projection operator in our original lists requires a tensor product of more than three H's, then we would have $M_{\text{H}} = 3$. Note that M_{H} depends *only* on the our quantum-assisted language and *not* on the particular program under consideration. Returning now to our simulation, since $\mathcal{P}_{\text{reg}}$ has traversed total $\mathcal{P}_{\text{quant}}$-difficulty $\mathcal{N}$ in each branch, it follows that the total number of operators that have been applied in each branch is bounded by $\mathcal{N}$, while the total number of H's that can occur in the tensor product in each branch is bounded by $M_{\text{H}}\mathcal{N}$. We have

$$f_{\text{reg}} \leq \{2^{\mathcal{N}}\}\ \{\mathcal{N} + \mathcal{N}^2\ m^{M_{\text{H}}\mathcal{N}}\}. \tag{16}$$

The first factor on the right is an upper bound on the total number of branches, where we are using the fact that each OBSERVE command spawns the splitting of one branch into two. The second factor on the right in (16) is an upper bound on the total $\mathcal{P}_{\text{reg}}$-difficulty of each branch. The first term therein covers the case in which the $\mathcal{P}_{\text{quant}}$-command simulated is INPUT, OUTPUT, APPEND, DELETE, IF, or APPLY. The second term covers the simulation of an OBSERVE command: The bound in this case is the product of our bound ($\mathcal{N}$) on the number of OBSERVE commands in a branch and the difficulty [from (13)] required to compute, for each OBSERVE command, the rational number to include in the special section. Combining (15) and (16), we obtain[18]

$$f_{\text{reg}} \leq a^{f_{\text{quant}}}, \tag{17}$$

[18]This equation appears to be nonsensical, in that it is not invariant under replacing f_{quant} by the equivalent difficulty function $2f_{\text{quant}}$. However, we broke this invariance in the derivation by assigning to each APPEND and OBSERVE command difficulty one.

where we have set $a = 4m^{2M_{\mathrm{H}}}$.

We conclude: Given any quantum-assisted program, there exists a regular program that computes exactly the same problem and has difficulty function satisfying (17). The benefit in efficiency from the quantum assist cannot be more than exponential. There is a more elegant, if slightly less informative, way of putting this. For f any difficulty function, denote by $\log f$ the difficulty function obtained by taking the log of f, possibly after adding to f a constant such that it is bounded away from 1. We note that the logs, so defined, of equivalent difficulty functions are equivalent. Then (17) implies that $\log f_{\mathrm{reg}} \leq f_{\mathrm{quant}}$. Of course, these general inequalities are rather coarse. If, in a particular example, a finer inequality is wanted, it usually can be obtained by applying (13) directly.

As an example of these ideas, let us return to the Grover construction. Here $m = 2$. Let us assign to each APPLY and OBSERVE command difficulty one, and suppose that none of these operators require a tensor product of more than two H's. Then $M_{\mathrm{op}} = 1$ and $M_{\mathrm{H}} = 2$. Choose, for $a > 2^{2M_{\mathrm{op}}} m^{2M_{\mathrm{H}}}$, the value $a = 64$.

Our quantum-assisted program computes this problem with difficulty function $f_{\mathrm{quant}}(n) = \sqrt{N}h(n)$, where $N = 2^n$. The inequality (17) now implies the existence of a regular program to compute this problem, with difficulty function $f_{\mathrm{reg}}(n) \leq (64)^{\sqrt{N}h(n)}$. We can find a much better bound than this. This particular program requires, for given n, just n H's in the tensor product, and it applies to this Hilbert space just $h(n)$ operators. Thus, to simulate a single OBSERVE requires of $\mathcal{P}_{\mathrm{reg}}$ difficulty $h(n)2^n$. There is a total of n such OBSERVE commands to be executed, and so we obtain $f_{\mathrm{reg}}(n) \leq nNh(n)$. Recall, by contrast, that the actual regular program for this construction has an even smaller difficulty function, namely, $f_{\mathrm{reg}}(n) = Nh(n)$. The extra factor of n in the former reflects the fact that our simulation recomputes $\gamma(\mathbf{S})$, from scratch, for each OBSERVation, whereas it is more efficient to carry out these n computations together.

Chapter 21

Quantum-Assisted Efficiency II

Can quantum-assisted programs offer any gain in efficiency over regular programs? Here is a precise formulation of this question.

Conjecture.[19] There exists a problem π, together with a quantum-assisted program $\mathcal{P}_{\text{quant}}$ (difficulty function f_{quant}), that computes π, such that there exists no regular program, $\mathcal{P}_{\text{reg}}$, that computes this same problem and whose difficulty function satisfies $f_{\text{reg}} \leq f_{\text{quant}}$.

As far as I am aware, we have neither a proof of nor a counterexample to this conjecture.

Note that, for a proof of the conjecture, one must actually *prove* (not merely suspect) that there exists no regular

[19] It is tempting to pose a stronger conjecture, asserting the existence of π and $\mathcal{P}_{\text{quant}}$ such that *every* regular program that computes this problem satisfies $f_{\text{quant}} \leq f_{\text{reg}}$ but not $f_{\text{quant}} \sim f_{\text{reg}}$. But this conjecture is very unlikely to be true (even though, as far as I am aware, we do not have a counterexample). The reason is that, given π and $\mathcal{P}_{\text{quant}}$, one can normally design a regular program that, although it may have considerably greater difficulty than $\mathcal{P}_{\text{quant}}$ for most input strings, is less difficult for an occasional string. We saw examples of this sort of construction in Sect. 12.

program at least as efficient as the given quantum-assisted one. What makes this conjecture hard is that we do not currently have good lower limits on the difficulty functions for the regular programs that compute a given problem. A good example is the prime problem: the problem of factoring a given integer n into its prime factors. There is no known method for computing this problem with a regular program whose difficulty function is $\leq (\log n)^{\mathrm{s}}$ for every positive number s. Yet, there are indications [12] that there exists, for this problem, a quantum-assisted program[20] whose difficulty function does satisfy this condition function $\ll n$. Thus, the prime problem is, arguably, a plausible candidate for an example as demanded by the conjecture. Surely, one might think, the most promising line to prove the conjecture would be to try to prove that this particular example works (i.e., that there is no regular program that computes the prime problem and whose difficulty function is $\leq (\log n)^{\mathrm{s}}$ for every $s > 0$). I would like to suggest, however, that this line may not be as promising as it first appears. There has been an enormous effort, by many talented people over many years, to prove that there is no easy computation of the prime problem. Yet this question remains open—and there is a good chance that it will remain open for some time to come. It may well be, in other words, that it is actually easier to settle the conjecture above than the question of the difficulty of the prime problem. But, of course, the prime problem is also of interest for other reasons.

We remark, below, on a few possible directions for proving (or disproving) this conjecture.

The obvious way to prove the conjecture would be to construct the problem π by a diagonal argument. That is, we would introduce the list of all quantum-assisted programs that compute problems, the list of all regular programs that compute problems, and the list of all input strings. To choose what π is on the first string, S_1, we would run a few quantum-

[20] However, it is not entirely clear whether such a program can be designed that meets all the requirements of Sect. 17 (in particular, that all operators are fixed initially, independent of the input string).

assisted programs on this string, as well as a few regular programs, determining, for these runs, what final strings result and what the difficulties are. Then, we would select $\pi(S_1)$ to eliminate the low-difficulty regular programs as well as the high-difficulty quantum-assisted programs. Continuing in this way through the list of input strings, we would hope to design a π with a quantum-assisted program that computes π, such that no regular program is at least that efficient. This appears to be a natural line (similar to the Blum [3] proof of Sect. 10). But despite its apparent promise this line has not so far met with success.

Another strategy involves exploiting probabilistic programs. We have seen, in Sect 19, that every probabilistic program may be simulated by a quantum-assisted program with the same difficulty function. Suppose, then, that we were able to find a probabilistic program whose difficulty function is unmatched by any regular program. This would yield immediately a proof of the conjecture. This strategy appears promising, for the language of probabilistic programs is somewhat simpler than that of quantum-assisted programs. Indeed, whereas quantum-assisted programs require additional objects (the Hilbert space H, the state $|\psi_0\rangle$, and some unitary and some projection operators on tensor products of H's), probabilistic programs do not. Whereas quantum-assisted programs require two additional commands (APPLY and OBSERVE), probabilistic programs do not. Whereas quantum-assisted programs require conditions on the function γ on history strings, probabilistic programs do not. The downside of this strategy is that quantum-assisted simulations of probabilistic programs do not appear to exploit what appears to be the key advantage of quantum-assistance: use of entanglements in the tensor product. In any case, as remarked in Sect. 13, there is no known example of a problem and a probabilistic computation of that problem for which one can prove that no regular program is at least as efficient.

Another class of possible examples comes from the Grover construction (Sects. 15 and 16). Suppose that we could find a suitable example of a needle-in-the-haystack problem. By "suitable," we mean one for which there is a regular program for checking needle candidates; but there exists no regular program that can find the needle any more efficiently than merely checking all possible candidates, one at a time. Then, as discussed in Sect. 16, this arrangement might lead to a problem and quantum-assisted program that satisfy the condition of the conjecture. But no such example, apparently, is known.

Another strategy is to try to construct, by using the quantum-assisted programming language itself, a problem for which quantum-assisted programs are well suited but regular programs are not. Consider, for example, the following: Let π accept as input a pair $(\mathcal{P}_{\text{quant}}, S_{\text{in}})$, where $\mathcal{P}_{\text{quant}}$ is a string representing a quantum assisted program and S_{in} is any string; and let π, on such input, return the string that is determined by running program $\mathcal{P}_{\text{quant}}$ on input string S_{in}. Quantum-assisted programs are certainly well set up for this π! But, unfortunately, this π is not even a problem, for $\mathcal{P}_{\text{quant}}$, applied to S_{in}, need not determine any string at all: There may be a nonzero probability that the program, on this string, will fail to halt altogether; or, even if it does halt, it may do so such that no one output has a probability strictly greater than that of every other possible output string. It would not help to modify this example to read as follows: $\pi(\mathcal{P}_{\text{quant}}, S_{\text{in}})$ is the empty string in case $\mathcal{P}_{\text{quant}}$, applied to S_{in}, fails to compute any string; otherwise it is whatever string $\mathcal{P}_{\text{quant}}$ does compute. Now we do indeed have a problem π but, unfortunately, it is not a computable one. Indeed, there exists no program that will even decide whether or not a given regular program and input string results in a halt.

More promising is to focus on the essence of quantum-assisted computing: the function γ. Fix a Hilbert space H, an initial state $|\psi_0\rangle$, and finite collections of unitary and of projection operators, each acting on some finite tensor product of H's. Recall, from Sect. 18, that a string $\mathbf{S}$ is called a *history* if it represents a finite, ordered, list of unitary or projection operators from this collection, where for each operator there is specified the particular copies of H on which it is to act, and for each projection operator there is assigned a result (0 or 1) of an observation via that operator. Fix a history string, $\mathbf{S}$. Then (i) assemble a tensor product of H's, consisting of those on which the operators in $\mathbf{S}$ act; (ii) consider the state $|\psi_0\rangle \cdots |\psi_0\rangle$ in this tensor product; (iii) apply to this state the operators of $\mathbf{S}$, in order, except that, for each projection $\mathcal{P}$ that is assigned result 0, apply $I - P$ rather than P; and (iv) take the squared norm of the resulting state. This is the number we denoted $\gamma(\mathbf{S})$ in Sect. 18, and we required there that it be rational-valued. In physical terms, $\gamma(\mathbf{S})$ is the probability that the sequence of operations and observations represented by $\mathbf{S}$ will in fact return the results we have assigned to each of the observations. This function γ contains all the information about quantum mechanics needed to simulate quantum-assisted programs written in this language.

Fix H, $|\psi_0\rangle$, and the collections of unitary and projection operators. Consider the problem γ itself [that is, the map that assigns to history string $\mathbf{S}$ the (rational) number $\gamma(\mathbf{S})$]. This is indeed a problem, and, by virtue of the restrictions we imposed in Sect. 18, it is a computable problem. If we could compute this problem γ by some relatively efficient regular program, then we could simulate each quantum-assisted program by a regular program with the same difficulty function and thus would conclude that the conjecture is false. But, as we remarked in Sect. 18, there is no obvious efficient method for computing γ with a regular program, and this observation is the essence of the idea that quantum mechanics might make for more efficient computation.

So, let us take, for the problem π of the conjecture, the problem γ itself. It is, as we already remarked, unlikely that this π can be computed efficiently by a regular program. However, this example does not seem to work either, for no quantum-assisted program can (at least not in any obvious way) do any better! Quantum-assisted programs are very good at taking actions in response to probability $\gamma(\mathbf{S})$

(for that is the essence of the OBSERVE command in that language), but they do not seem particularly adept at actually computing the integers that appear in the numerator and the denominator of this fraction.

Here is a more promising way to incorporate this γ into a problem. Let, for $\mathbf{S}$ any history string, $\pi(\mathbf{S})$ be "yes" if $\gamma(\mathbf{S}) \geq 1/2$ and "no" if $\gamma(\mathbf{S}) < 1/2$. This is, again, a computable problem, and, again, it is plausible that there is no regular program that computes it efficiently. But here is a relatively efficient quantum-assisted program $\mathcal{P}_{\text{quant}}$ for this problem. Let $\mathcal{P}_{\text{quant}}$, given a history $\mathbf{S}$, simply perform, physically, the operations represented by $\mathbf{S}$, keeping a record of the results of any OBSERVE commands. Then, $\mathcal{P}_{\text{quant}}$ compares those actual results with the results already encoded in $\mathbf{S}$ and reports yes if they agree and no if they do not. This quantum-assisted program $\mathcal{P}_{\text{quant}}$ computes this problem π, except for one little thing. For any string for which $\gamma(\mathbf{S})$ has exactly the value 1/2, then this $\mathcal{P}_{\text{quant}}$ computes nothing (for it will return yes or no, each with probability 1/2). But this is easily remedied by slightly modifying $\mathcal{P}_{\text{quant}}$. First note that, if history string $\mathbf{S}$ contains exactly L projection operations, then $\gamma(\mathbf{S})$ is a fraction with denominator 2^L. Thus, we have only to modify $\mathcal{P}_{\text{quant}}$ so that, for input string $\mathbf{S}$, it reports "yes" with probability $\gamma(S) + 1/2^{L+1}$ [rather than just $\gamma(\mathbf{S})$ as before] and no otherwise. This is easily accomplished by incorporating into $\mathcal{P}_{\text{quant}}$ a quantum-generated additional probability of $1/2^{L+1}$ for "yes".

So, we have a quantum-assisted program $\mathcal{P}_{\text{quant}}$ that computes the problem π. Note that this $\mathcal{P}_{\text{quant}}$ is relatively efficient when $\gamma(\mathbf{S})$ is far from the value 1/2—say, less than 1/3 or more than 2/3. However, when $\gamma(\mathbf{S})$ is close to 1/2—and, in particular, when it is precisely 1/2—then $\mathcal{P}_{\text{quant}}$ will be very inefficient, a consequence of the fact that, since the numbers of yes and no answers will be nearly equal, many runs of $\mathcal{P}_{\text{quant}}$ will be necessary to determine $\pi(\mathbf{S})$. Indeed, when $\gamma(\mathbf{S})$ is close to 1/2, $\mathcal{P}_{\text{quant}}$ may be less efficient than the regular program that computes π. We could (although it is not necessary, in light of the way the conjecture is structured) adjust for this by modifying $\mathcal{P}_{\text{quant}}$ further. While running its quantum-assisted computation of $\pi(\mathbf{S})$, $\mathcal{P}_{\text{quant}}$ also carries out the regular computation of $\gamma(\mathbf{S})$. Then, $\mathcal{P}_{\text{quant}}$ reports whichever method terminates first.

So, we have a problem π and a quantum-assisted program $\mathcal{P}_{\text{quant}}$ that computes π, such that the obvious regular program to compute π is not more efficient that $\mathcal{P}_{\text{quant}}$. But this alone does not establish the conjecture: We must *prove* that there exists *no regular program whatever* more efficient that $\mathcal{P}_{\text{quant}}$. But—and this will come as no surprise—obtaining such a proof does not seem to be an easy task.

We remark that there are some pretty tough looking subproblems of this problem π. For example, let the Hilbert space H be two dimensional, and let the unitary operators include a "spin-flip" operator (that reverses up-spin and down-spin) and a Toffoli operator (that flips one spin if and only if two others are up). Let there be just one projection operator, the "spin-up" projection. Consider history strings that (i) create a tensor product of n copies of H (so the Hilbert space of states has dimension 2^n), (ii) create some initial state in this Hilbert space, (iii) apply to this state m Toffoli operators, acting on various of the factors, and (iv) OBSERVE the spin-state in the first H-factor. Now, for many such history strings, $\gamma(\mathbf{S})$ will be close to 1/2; but there will also be many for which $\gamma(\mathbf{S})$ is far from 1/2. For the latter, quantum-assisted program $\mathcal{P}_{\text{quant}}$ will compute $\pi(\mathbf{S})$ with difficulty $n + m$. It appears that it will be extremely difficult to find a regular program that will be more efficient for these history strings. Thus, it appears plausible that this π and $\mathcal{P}_{\text{quant}}$ will satisfy the condition of the conjecture.

In summary, there are a number promising-looking strategies one might employ to try to decide whether the conjecture above is true or false, but none, so far, has panned out. It appears that the question posed by this conjecture is a difficult one.

Chapter 22

Conclusion

We have discussed here three broad aspects of the theory of computation.

The first of these is the notion of computability—what can, in principle, be computed. This subject is, by any measure, in excellent shape. There is apparently a unique, natural notion of what it means to "compute" something. And we can produce simple examples of problems that are computable and of those that are not.

The second aspect involves the notion of the difficulty of a computation—roughly speaking, the number of steps required to carry it out. We introduce what purports to be "the simplest efficient language" and, by means of it, define what we mean by a "method" to compute a problem, as well as by the difficulty of that method. It has been proved that there exist "very hard problems", and that there exist problems for which there is no "most efficient" computation. Furthermore, there is a simple, natural way (by merely altering slightly one of the commands of this language) to incorporate probability into the computation process. A probabilistic program returns, for a given input string, an "answer" only probabilistically. Nevertheless, there is a suitable definition of what it means for such a program to compute a problem, and one can assign, in a natural way, a difficulty function to such a computation. There are at least three open issues in this area. First, although our definition of difficulty is perhaps reasonable, there do exist some technical variants of it, and there is no solid argument that our scheme is "more reasonable" than these alternatives. Can such an argument be found? Second, it turns out that, for virtually every interesting problem, we do not have a good lower bound on the difficulty of the regular computations of that problem. This lack of good lower bounds is arguably *the* outstanding gap in this subject, and an enormous amount of effort has gone into it. And, finally, it is not known whether there exists a problem, along with a probabilistic computation of that problem, such that no regular computation is equally efficient. It is surprising, to say the least, that such a question should remain open. It would be most interesting to settle it.

The third aspect involves the use of quantum mechanics in the computation process. It turns out that one can introduce a certain, precise computer language, designed to reflect what (and only what) could be done, in the laboratory, using quantum systems. This language introduces "basic quantum systems," with a finite-dimensional Hilbert space, unitary evolution, and certain observables. Computations are carried out using tensor products of copies of these basic systems. One introduces, further, a suitable difficulty function for computations in this language, designed to reflect the physical difficulty of an actual computation. This, the language of "quantum-assisted computing," allows us to reformulate questions about physical computers into questions about mathematics. Using this formulation, for example, we can prove that, given any computation of a problem using quantum mechanics, there is a computation that does not use quantum mechanics, that is at most exponentially more difficult. A central question is whether or not there exists a problem, together with a quantum-assisted computation of that problem, such that that computation is more efficient than *any* nonquantum computation. Although there certainly are indications that there does exist such a problem, we have today neither a proof nor a counterexample. This, to my mind, is one of the most fascinating questions in the subject of quantum-assisted computing.

Quantum mechanics brings to the computation process two quite separate potential advantages. One involves the use of probability. We can write quantum-assisted programs that, essentially, use quantum mechanics only as a random number generator (i.e., that merely mimic probabilistic programs). But, even when restricted to this aspect, quantum mechanics has the potential to enhance the computation process: It is an open question whether there exists a probabilistic program (and, therefore, a quantum-assisted program invoking only random number generation) that cannot be matched, in terms of efficiency, by some regular program. The second involves the use of entanglements, that is, of the ability, with quantum mechanics, to manipulate large numbers of terms in a tensor product in a single step. It is from this source that the advantages of quantum mechanics—if there are any at all—are likely to be the most dramatic. These two appear to be quite separate effects, relying on very different features of quantum mechanics. In light of all this, it is strange that the example in which the probability aspect of quantum mechanics might play a role (Sect. 13), and that in which entanglements might play a role (Sect. 15) are strikingly similar. Is it possible that, on some deeper level, these two aspects are somehow related to each other?

It would be interesting to try to do with classical mechanics what has been done with quantum mechanics. That is, we would introduce a precise mathematical language of "classical-mechanics-assisted computing," designed to reflect how classical mechanics might be used in the laboratory. With such a language in hand, we could, for example, ask whether or not there exist a problem and a classical-mechanics-assisted computation of that problem, such that the efficiency of that computation cannot be matched with any regular program. Let us, for example, posit that "classical mechanics" is to be idealized as follows: A system is to be described by a manifold of states (e.g., its phase space) together with a dynamical vector field on that manifold. The integral curves of the dynamical vector field give the evolution of the system through time. We identify certain regions of this manifold as corresponding to the various input strings and certain other regions to various output strings. This arrangement may now be used to "compute" in the following manner: Given any input string, begin with the system in a state lying in the corresponding input-string region of this manifold. Then evolve (i.e., follow the dynamical vector field) until we arrive at some output-string region. It turns out, unfortunately, that these particular rules for classical-mechanics-assisted computing are too permissive. It is, for example, not difficult to specify a particular manifold, along with a vector field and such regions, such that the resulting system, under these rules, computes the halting problem. We have merely to encode, into the vector field and the regions, which programs halt and which do not. Thus, the problem with the framework above is that it does not impose, on the vector field and region assignments, some suitable requirement to the effect that they be "physically constructible." It is very hard to think of any mathematical condition that could be imposed on the manifold, the vector field, and the regions that would reflect such a requirement.

A similar situation can arise already in the quantum case. Let us idealize quantum mechanics as follows: A quantum system is described by a Hilbert space together with a family of unitary operators, giving the time-evolution, and also with a collection of self-adjoint operators, giving the observables. These observables are to be interpreted as representing the input and output strings. This arrangement could be is used to compute, in a manner similar to that of classical mechanics. Start the system in an appropriate eigenstate of an input-string observable, evolve the system using the unitary operators, and make a final observation via an output-string observables. But, just as with classical mechanics, this arrangement could be used to compute the halting problem. However, there is one crucial difference, in this regard, between classical and quantum mechanics: Whereas it appears very difficult, in the case of classical mechanics, to invent new rules that can be imposed to prevent this sort of thing, it is relatively easy to do so in quantum mechanics. We first introduce a very simple quantum system, consisting of a finite-dimensional Hilbert space with a couple of simple unitary and self-adjoint operators thereon. This system is relatively structureless—it is not rich enough to encode, for example, the solution to the halting problem. We now build our quantum-assisted computer by taking (finite) tensor products of copies of this simple system. In other words, we demand that the quantum system we use for our computation be explicitly constructed from these simple building blocks. In this way, we are able to use quantum mechanics to assist in the (complicated) computation of a problem, without the danger of encoding the solution of the problem, right from the beginning, in the quantum system itself. It is difficult to think of any way to do a similar thing for classical mechanics. What, for example, are the analogous building blocks?

It is the tensor product that gives quantum mechanics its potential advantage in efficient computing. The tensor product of n physical systems, each having, say, m states, describes a system whose general state is a superposition of m^n states. We thus can, by relatively simple manipulations (i.e., applying operators to one or two of the n systems), manipulate m^n numbers (the coefficients in the superposition). Can we find other physical theories that might employ a similar advantage? That is, do we find, in any other physical theories, a "tensor-product-like" construction?

Consider electromagnetism. Suppose that we were capable of manufacturing small boxes, in which there could be stimulated a total of three possible electromagnetic modes. Thus, the electromagnetic states within each box form a three-dimensional (real) vector space, V. Now take two such boxes, place them side by side, and regard these two as a single system. What is the space of states of this combination? Well, each of the two boxes carries a field, in some state in V, and so the state of the total system is described by simply specifying these two elements of V. That is, the vector space of states is $V \oplus V$, the direct sum of V and V (with dimension $6 = 3 + 3$). Had this instead been $V \otimes V$, the tensor product (with dimension $9 = 3 \times 3$), then we would have the beginnings of a promising theory of electromagnetic-assisted computing.

Tensor products, it appears, do not routinely make an appearance outside of quantum mechanics. Is there some general principle of nonquantum physics that rules out the tensor product, once and for all? The following example may shed some light on this question. Consider a one-dimensional "box," of length L, in which the vector space V of allowed states is that resulting from exciting three modes of a field, given, say, by ($\sin \pi x/L$, $\sin 2\pi x/L$, $\sin 3\pi x/L$). Here, for this example, is a mechanism to realize the tensor product, $V \otimes V$. Consider fields in the square of side L. The corresponding modes are arbitrary linear combinations of products, $\sin a\pi x/L \sin b\pi y/L$, where $a, b = 1, 2, 3$. The vector space of such solutions is indeed the nine-dimensional $V \otimes V$. Similarly, passing to a cubic box, we obtain a space of field states given by $V \otimes V \otimes V$. Here, in other words, is a situation in which we *can*, physically, form tensor-product states. But just this three-fold tensor product is not good enough—we must be able to take arbitrarily large tensor products if this scheme is likely to be useful in computations. Alas, we all too soon run out of dimensions.

It might also be of interest to try to characterize all problems (or, at least, some large class of problems) for which quantum-assisted computation is more efficient than computation without quantum mechanics (assuming, for the moment, that there exist any such problems at all!). Even the question itself must be stated with care, in light of the dependence of computational efficiency on the method employed. We might like to ask, for example, whether a given quantum-assisted program for computing a problem is more efficient that the nonquantum program "using the same method." But, unfortunately, we do not at the moment have any notion of "same method" for quantum-assisted and nonquantum-assisted programs. Can we find a nontrivial class of problems for which we can *prove* that, in some suitable sense, for no problem in this class does quantum assist offer any advantage?

Finally, it might be interesting to understand in some deeper sense how our physical theories interact with the mathematics of computation. Can one, for example, imagine a plausible-looking physical theory within whose framework certain computations can be speeded up even more dramatically?

References

[1] Agrawal, M, Neerja, K, and Nitin, S, "Primes Is in P," Ann. Math. 160, 781 (2004). Available at http://www.math.princeton.edu/~annals/issues/2004/Sept2004/Agrawal.pdf. This paper shows that there exists a program that computes the problem of deciding whether or not an integer n is prime, with difficulty function $\leq (\log n)^s$, for every $s > 15/2$.

[2] Arora, S., and Barak, B, 'Complexity Theory.' Available at http://www.cs.princeton.edu/theory/complexity/. This is a readable summary of (mostly nonquantum) complexity theory.

[3] Blum, M, "A Machine-Independent Theory of the Complexity of Recursive Functions," J. Assoc. Comput. Mach. 14, 322 (1967). Available at http://portal.acm.org/citation.cfm?coll=GUIDE&dl=GUIDE&id=321395. This paper deals with properties of difficulty that rely only on some general features of the difficulty-measure. There are two very nice results here.

[4] Grover, L, "A Fast Quantum-Mechanical Algorithm for Database Search," in Proc. 28th Annual ACM Symposium on Theory of Computing, ACM, New York (1996).

[5] Hartmanis, J, and Hopcroft J. E., "An Overview of the Theory of Computational Complexity," J. Assoc. Comput. Mach. 18, 444 (1971).

[6] Hennie, F. C., "One-Tape Off-Line Turing Machine Computations," Inform. Control 8, 553 (1965). It is proved that no Turing machine can compute the palindrome problem with difficulty function $\ll L(S)^2$.

[7] Karatsuba, A, and Ofman, Y, "Multiplication of Many-Digital Numbers by Automatic Computers," Doklady Akad. Nauk SSSR 145, 293 (1962); translation in Phys. Doklady 7, 595 (1963). This paper presents a method to multiply two n-digit numbers with difficulty function $\ll n^2$.

[8] Kelly, J., *General Topology*, Springer-Verlag, New York (1975). An appendix contains the best treatment of axiomatic set theory I have ever seen.

[9] Lenstra, A. K., and Lenstra, H. W, eds., *The Development of the Number-Field Sieve*, Lecture Notes in Mathematics 1554, pp. 11–42, Springer-Verlag, New York (1993). The sieve method for computing the prime problem is described here.

[10] Mermin, D., "Quantum Computation," lecture notes. Available at http://people.ccmr.cornell.edu/~mermin/qcomp/CS483.html. These lecture note are a nice elementary introduction to this subject. See especially Sect, IV.

[11] Pittenger, A. O., *An Introduction to Quantum Computing Algorithms*, Progress in Computer Science and Applied Logic, Vol. 19, Birkhauser, Boston (2001). This book gives some detail on how to do real-world computations with c-not gates.

[12] Shor, P. W, "Polynomial-Time Algorithms for Prime Factorization and Discrete Logarithms on a Quantum Computer," SIAM J. Sci. Stat. Comput. 26, 1484 (1997). This paper proposes a method for using quantum mechanics for efficient factorizing of integers.

[13] Unruh, W. G., "Maintaining Coherence in Quantum Computers," Phys. Rev. A 51, 992 (1995).

[14] Yasuhara, A., *Recursive Function Theory and Logic*, Academic Press, San Diego (1971). This is my favorite book on Turing machines, unsolvable problems, etc.

Index